Health Care Engineering: Part II

Research and Development in the Health Care Environment

Synthesis Lectures on Biomedical Engineering

Editor

John D. Enderle, *University of Connecticut*

Lectures in Biomedical Engineering will be comprised of 75- to 150-page publications on advanced and state-of-the-art topics that spans the field of biomedical engineering, from the atom and molecule to large diagnostic equipment. Each lecture covers, for that topic, the fundamental principles in a unified manner, develops underlying concepts needed for sequential material, and progresses to more advanced topics. Computer software and multimedia, when appropriate and available, is included for simulation, computation, visualization and design. The authors selected to write the lectures are leading experts on the subject who have extensive background in theory, application and design.

The series is designed to meet the demands of the 21st century technology and the rapid advancements in the all-encompassing field of biomedical engineering that includes biochemical, biomaterials, biomechanics, bioinstrumentation, physiological modeling, biosignal processing, bioinformatics, biocomplexity, medical and molecular imaging, rehabilitation engineering, biomimetic nano-electrokinetics, biosensors, biotechnology, clinical engineering, biomedical devices, drug discovery and delivery systems, tissue engineering, proteomics, functional genomics, molecular and cellular engineering.

Health Care Engineering: Part II: Research and Development in the Health Care Environment
Monique Frize
November 2013

Health Care Engineering: Part I: Clinical Engineering and Technology Management
Monique Frize
November 2013

Computer-aided Detection of Architectural Distortion in Prior Mammograms of Interval Cancer
Shantanu Banik, Rangaraj M. Rangayyan, J.E. Leo Desautels
January 2013

Fundamentals of Biomedical Transport Processes
Gerald E. Miller
2010

Models of Horizontal Eye Movements, Part II: A 3rd Order Linear Saccade Model
John D. Enderle andWei Zhou
2010

Models of Horizontal Eye Movements, Part I: Early Models of Saccades and Smooth Pursuit
John D. Enderle
2010

The Graph Theoretical Approach in Brain Functional Networks: Theory and Applications
Fabrizio De Vico Fallani and Fabio Babiloni
2010

Biomedical Technology Assessment: The 3Q Method
PhillipWeinfurt
2010

Strategic Health Technology Incorporation
BinsengWang
2009

Phonocardiography Signal Processing
Abbas K. Abbas and Rasha Bassam
2009

Introduction to Biomedical Engineering: Biomechanics and Bioelectricity - Part II
Douglas A. Christensen
2009

Introduction to Biomedical Engineering: Biomechanics and Bioelectricity - Part I
Douglas A. Christensen
2009

Landmarking and Segmentation of 3D CT Images
Shantanu Banik, Rangaraj M. Rangayyan, and Graham S. Boag
2009

Basic Feedback Controls in Biomedicine
Charles S. Lessard
2009

Health Care Engineering: Part II: Research and Development in the Health Care Environment
Monique Frize, P. Eng., O.C.

www.morganclaypool.com

ISBN: 9781627050722 print
ISBN: 9781627050739 ebook

DOI 10.2200/S00540ED1V01Y201310BME051

A Publication in the Morgan & Claypool Publishers series
SYNTHESIS LECTURES ON BIOMEDICAL ENGINEERING #51
Series Editor: John D. Enderle, University of Connecticut

Series ISSN 1930-0328 Print 1930-0336 Electronic

Health Care Engineering: Part II

Research and Development in the Health Care Environment

Monique Frize
Carleton University and University of Ottawa

SYNTHESIS LECTURES ON BIOMEDICAL ENGINEERING #51

ABSTRACT

Chapter 7 presents some statistics on the occurrence of medical errors and adverse events, and includes some technological solutions. A chapter on electronic medical records follows. The knowledge management process divided into four steps is described; this includes a discussion on data acquisition, storage, and retrieval. The next two chapters discuss the other three steps of the knowledge management process (knowledge discovery, knowledge translation, knowledge integration and sharing). The last chapter briefly discusses usability studies and clinical trials.

The two parts consolidate material that supports courses on technology development and management issues in health care institutions. It can be useful for anyone involved in design, development, or research, whether in industry, hospitals, or government.

KEYWORDS

medical errors and adverse events, information technologies, electronic medical records, knowledge management process, clinical trials, usability studies

Contents

Preface

The material for this Lecture Series arises from twenty years of teaching a course entitled "Health Care Engineering." My background was an undergraduate degree in Electrical Engineering at the University of Ottawa, a Master of Philosophy in Electrical Engineering in Medicine at Imperial College of Science and Technology in London (U.K.) and a doctorate at Erasmus Universiteit in Rotterdam (The Netherlands) in clinical engineering. After the Master degree in London in 1970, I taught two courses—bioinstrumentation and biomedical statistics—at Université du Québec à Montréal for one term, then worked as a clinical engineer at Notre-Dame Hospital for eight years. In this role, I developed the medical equipment management program and realized quickly how engineers have a significant impact on the quality, effectiveness and safety of health care delivery. In July 1979, I was appointed Head of the Regional Clinical Engineering Services for seven hospitals in North-Eastern New-Brunswick. In 1989, after completing the doctoral degree, I was appointed Professor in Electrical/Biomedical Engineering at the University of New Brunswick, where I developed the health care engineering course (Part I). In 1996, as Professor at Carleton University and the University of Ottawa, I developed a second course for graduate students (Part II).

This book provides a basis on which other specialized courses can be built, such as signal processing, medical imaging, data mining, processing, etc.. It is useful for hospital administrators, clinical engineers, and biomedical technicians who hold a direct or indirect role for technology management in their institution. References and additional readings are provided for readers who wish to have a more in-depth knowledge about each of the topics discussed.

This book is Part II of a two-part series on health care engineering. Chapter 7 presents a brief discussion on the occurrence of adverse events and medical errors and how information technology can help to reduce these. Chapter 8 discusses issues related to electronic medical records. The next three chapters present the four steps in the knowledge management process and illustrate these through examples taken from the perinatal clinical environment. These steps include data acquisition, storage, and retrieval; knowledge discovery; knowledge translation; and knowledge integration and sharing. Chapter 12 concludes the book with a short discussion on clinical trials and usability studies.

This book consolidates material that supports a course on medical technology management and research and development of technologies in a health care environment. It can be useful for anyone involved in design, development, or research, whether in industry, hospitals, government, or in universities and colleges. It is not intended to cover all topics in depth but rather to provide an overview of the subjects and sources where additional information can be found.

I am grateful to all the students who have provided interesting discussions in their essays and class discussions. They have contributed positively to an improvement of the course over the years. My hope is that they become responsible engineers, who design and develop technological solutions with people, society, and our world in mind.

For those interested in Part I, it briefly describes the health care systems in Canada and the U.S. Chapter 2 discusses the measurement of physiological variables in humans, for the purpose of diagnosis or to monitor disease conditions or treatment effectiveness. Chapter 3 discusses the function and role of the clinical engineer in managing medical technologies in industrialized and developing countries. Safety considerations are discussed in Chapter 4 as the safe and effective use of technology. Chapter 4 also goes on to describe how to minimize liability and establish a quality assurance program for technology management. Telemedicine and some of its applications are presented in Chapter 5 and the technology assessment process is discussed in Chapter 6. A discussion on research methods can be found in *Ethics for Bioengineers.* by Monique Frize, Morgan and Claypool, 2011.

CHAPTER 7

Adverse Events, Medical Errors, and the Role of Information Technology in Reducing Them

7.1 SOME STATISTICS ON MEDICAL ERRORS AND ADVERSE EVENTS

The famous report, *To Err is Human*, defined medical errors as "the failure of a planned action to be completed as intended or the use of a wrong plan to achieve an aim" (Institute of Medicine, 1999). The report lists problems that commonly occur during the course of providing health care; examples are: adverse drug events and improper transfusions, surgical injuries and wrong-site surgery, suicides, restraint-related injuries or death, falls, burns, pressure ulcers, and mistaken patient identities. The report adds that high error rates with serious consequences are most likely to occur in intensive care units, operating rooms, and emergency departments. In the same report, an adverse event (AE) is defined as an injury caused by medical management rather than by the underlying condition of the patient (Institute of Medicine, 1999). The report claimed that, in the U.S., at least 44,000 people and perhaps as many as 98,000 people die in hospitals each year as a result of medical errors that could have been prevented. In addition to the loss of life, adverse events and medical errors may lead to hospitalization or to a prolonged stay; where a disability is suffered there can be a loss of income and a lower household productivity that range between $17 and $29 billion per year (1999 dollars) nationwide. There is also a loss of trust by patients in the health care system, and diminished satisfaction by both patients and health professionals. Long hospital stays or disability can result in physical and psychological discomfort; health professionals may suffer a loss of morale and frustration at not being able to provide the best care possible. For society, there can be a loss of worker productivity, reduced school attendance by children, and a lower overall level of population health.

An Australian report (Wilson et al., 1999) claimed that 34.6% of complications were due to the failure or the technical performance of an indicated procedure or operation; 15.8% were due to the failure to synthesize, decide, and/or act on available information; 11.8% were due to the failure to request or arrange an investigation, procedure, or consultation; and 10.9% were due to lack of care and attention or failure to attend the patient (Wilson et al., 1999). The report also mentioned that cognitive failure appears to have had a role in 57% of all adverse events (AE); most reported AEs were found to be highly preventable and were deemed to have caused significant disability; they were largely errors of omission rather than of commission. The authors argued there is a need to look at factors in health care delivery that interfere with the cognitive or technical performance of providers. The factors they listed that explain the occurrence of AEs and errors are: the insufficient use of information technologies to assemble information for decision-making, fatigue, sleep deprivation, level of supervision of junior staff, and a culture that portrays error as individual failure; a deviation from perfection was also perceived as another factor. Quoting these authors:

> Improvement is needed in the agreed processes of care, supported by information systems that allow general dissemination of current knowledge of diseases or treatments, and information on outcomes of care for each patient, through appropriate quality processes (**Wilson et al., 1999**).

Some examples suggested were: practice guidelines and protocols at the point-of-care; use of automated reminders for patients and providers when a test or a follow-up is required; and having adequate patient outcome information that can be benchmarked to identify unacceptable variations (Wilson et al., 1999).

A Canadian study looking at patient charts found 7.5% of reported AEs, of which 36.9% were preventable. Death had occurred in 20.8% of these cases and there were 1521 additional hospital days recorded. The cases concerned mostly older patients; the mean age was 64.9 years. In this study, there were 2.5 million hospital admissions and an estimated 185,000 AEs, with 70,000 of these being preventable. There is an important distinction between medical errors and adverse events; not all errors lead to an adverse event. It is the latter that causes harm to patients and sometimes death or disability. However, the prevention of errors is bound to have a positive effect on reducing the occurrence of adverse events (Baker et al., 2004).

7.2   TYPES OF MEDICAL ERRORS

Medical errors can be classified in different ways. One approach classifies medical errors in the following categories: medication, surgical, diagnostic, equipment-related, or a systems problem (Rideout, 2006). Some of the risk factors related to medication errors are: medications with the same name, or with a similar packaging, illegible physician handwriting, or a failure to include a leading zero. Surgical errors can arise when there is more than one surgeon, there are multiple procedures to be done on the same patient concurrently, or the surgical technique may be unfamiliar to the surgical team. For errors of diagnostic, there may be a delay in testing or in obtaining laboratory or imaging results or errors may arise from the fact that the diagnostic technique is not familiar to the clinical team. Errors related to the use of equipment may be due to defective components, poor construction or design, the presence of electromagnetic interference, an incorrect equipment set-up, adjustment or cleaning, or an operator error due to lack of proper training; there may also be devices with the same function, but from different manufacturers, therefore they may have several operating features that vary from one model to the other. Other risk factors are: length-of-stay in hospitals, too much reliance on memory, distraction or interruption of caregivers, high-pressure situations, and communication problems (Rideout, 2006).

Leape et al. (1993) classified error types as follows:

1. **Diagnostic**. A delay in reaching a diagnosis or an error in the diagnosis made by the physician. This includes a failure to employ indicated tests, or the use of outmoded tests or therapy, or a failure to act on results of monitoring or testing.

2. **Treatment**. Defined as an error in the performance of an operation, procedure, or test; error in administering the treatment; error in the dose or method of using a drug; avoidable delay in treatment or in responding to an abnormal test; and inappropriate or not indicated care.

3. **Preventive**. Would be a failure to provide prophylactic treatment, inadequate monitoring, or a follow-up of treatment. Other types would be a failure of communication, an equipment failure, or a system failure (Leape et al., 1993).

7.3   INFORMATION TECHNOLOGIES THAT CAN HELP REDUCE ERRORS AND ADVERSE EVENTS

There are a variety of information technologies that can help reduce the advent of medical errors or adverse events. Among these we find adverse event tracking systems and online clinical information systems using handheld computers (Fisher et al., 2003; Horsley and Forster, 2005). There exist pharmacy and nursing information systems and computerized physician order entry systems for medication management; there are also electronic medical records, automatic and biometric identification technologies, and clinical decision support systems. Other useful approaches are human factors engineering and usability testing to ensure that new technologies and procedures are appropriately tested for their ease-of-use and usefulness. (More discussion on testing can be found in Chapter 12.)

Bates and Gawande (2003) present a fairly comprehensive view on ways that information technology (IT) can reduce errors. They claim that IT can reduce the frequency of errors of different types and probably the frequency of adverse events. These authors claim that the main strategies to prevent errors and adverse events is to use tools that improve communication, make knowledge more easily accessible, assist with calculations, perform checks in real-time, assist with monitoring, and provide some decision-support tools.

Poor communication during handoff between physicians (a shift change) appears to be a common factor contributing to the occurrence of adverse events. The authors cite one study where the risk of adverse events increased by a factor of 5.2 in the cross-coverage of medical inpatients; the solution suggested was to use a computerized coverage system for signing out, using hand-held personal digital assistants, and having wireless access to electronic medical records to improve the exchange of information; this was especially effective if links between various applications and a common clinical database were in place, since many errors result from inadequate access to clinical data. Another suggestion arising from the study was that laboratory results, which are critical but rarely occur, can be buried with all other results; such results need to be identified and communicated quickly and automatically to clinicians, not just to a clerk in the unit. Access to drug reference, infectious disease information, and articles in Medline at the point-of-care is increasingly feasible with small hand-held devices (Bates and Gawande, 2003; Fisher et al., 2003).

Bates et al. (2001) made some general recommendations to reduce errors: implementing clinical decision support judiciously; considering consequent actions when designing systems; testing existing systems to ensure they actually catch errors that injure patients; promoting the adoption of standards for data and systems; developing systems that communicate with each other; using systems in new ways; measuring and preventing adverse consequences; making existing quality structures meaningful; and improving regulation and removing disincentives for vendors to provide clinical decision support. Specific recommendations were to implement provider order entry systems, especially computerized prescribing; to implement bar-coding for medications, blood, devices, and patients; and to utilize modern electronic systems to communicate key pieces of asynchronous data such as markedly abnormal laboratory values. The authors concluded that appropriate increases in the use of information technology in health care, especially the introduction of clinical decision support and better linkages in and among systems, resulting in process simplification, could result in substantial improvement in patient safety (Bates et al., 2001).

Assisting with calculations of infusion rates for pumps and controllers, and with weight-based calculations for medication doses and route administration can prevent serious errors. Allergies and drug interaction information can also prevent errors. Forcing functions were found to be one of the primary ways in which computerized order entry by physicians reduces the rate of errors. Bar-coded patient identification bracelets were also found to be helpful in eliminating errors, especially if they identify procedures intended for a particular patient (Bates and Gawande, 2003).

Another study of the technology-enabled remote monitoring of a 10-bed intensive care unit was associated with a reduction of mortality of 68% and 46% in two different time periods and the average length of stay and related costs decreased by about one third (Bates and Gawande, 2003).

Regarding decision-support, neural networks have been applied to reduce diagnostic or treatment errors with applications in the assessment of abdominal pain, chest pain, psychiatric emergencies, and in the interpretation of radiologic images and tissue specimens. This is discussed in more detail in Chapter 10 on Knowledge Discovery and data analysis tools and Chapter 11 on decision support tools.

REFERENCES

Baker, GR, Norton, PG, Flintof, V, Blais, R, Brown, A, Cox, J, et al. (2004). "The Canadian Adverse Event Study: the incidence of adverse events among hospital patients in Canada." *CMAJ*, May 25, 170(11): 1678-1686.

Bates, DW, Cohen, M, Leape, LL, Overage, JM, Shabot, MM, Sheridan, T (2001). "Reducing the Frequency of Errors in Medicine Using Information Technology." *JAMIA*, 8(4): 299-308. DOI: 10.1136/jamia.2001.0080299.

Bates, DW, Gawande, AA (2003). "Improving Safety with Information Technology." *NEJM*, 348: 2526-2534. DOI: 10.1056/NEJMsa020847.

Fisher, S, Stewart, TE, Mehta, S, Wax, R, Lapinsky, SE (2003). "Handheld Computing in Medicine." *J. Am. Med. Inform. Assoc.* (*JAMIA*), 10:139-149. DOI: 10.1197/jamia.M1180.

Horsley, A, Forster, L (2005). "Handheld Computers in Medicine: The Way Forward. A Short Report." *Postgrad. Med. J.*, 81:481-482. DOI: 10.1136/pgmj.2004.029793.

Institute of Medicine (1999). "To Err is Human: Building a Safer Health System." (A brief version of the report is available at: http://www.iom.edu/~/media/Files/Report%20Files/1999/To-Err-is-Human/To%20Err%20is%20Human%201999%20%20report%20brief.pdf; last accessed July 2013.

Leape, L, Lawthers, AG, Brennan, TA, et al. (1993). "Preventing Medical Injury." *Qual. Rev. Bull.*, 19(5): 144–149.

Rideout, K (2006). "Identification of Primary Risk Factors of Adverse Medical Events Using Artificial Neural Networks." MASc thesis, Systems and Computer Engineering, Carleton University, Ottawa, Canada K1S 5B6.

Wilson, H., Gibberd, H (1999). "An Analysis of the Causes of Adverse Events from the Quality in Australian Health Care Study." *Med. J. Aust.*, 170: 411-415.

OTHER SUGGESTED READING

Bogner, MS, Ed. (1994). *Human Error in Medicine*. Lawrence Erlbaum Associates, Publishers, Hillsdale, NJ.

CHAPTER 8

The Electronic Medical Record (EMR): Design, Safety, and Meaningful Use

This chapter presents a brief historical development of the EMR and the benefits of a wide adoption of EMRs by physicians in hospitals, health centers, and private practice. Current literature shows that the rate of adoption of the technology varies greatly between countries; several barriers act as a disincentive to their use, but recent government policies have encouraged their adoption. Currently, there are many companies offering this product, so it is important to decide what design characteristics are important to ensure a meaningful use of the EMR by physicians, nurse practitioners, and other caregivers who need to access these records. There are safety concerns to consider for a successful implementation. Another major debate concerns patient access to the EMR, and who owns the information stored in it. The chapter ends with suggestions regarding the design of EMRs. The quotes below show some of the benefits of using EMRs.

"Imagine walking into a clinic to see a doctor you have not met before. With your authorization, the doctor can obtain your essential health care information, such as medications, X-rays and the names of the other clinicians who have cared for you..." (**Canada Health Infoway**)

and

"Imagine you are the doctor. With this information, you will have a better understanding of your new patient's health care needs and be better equipped to make informed decisions about treatment and subsequent self-management of the patient's own health." (**Canada Health Infoway**)

8.1 WHAT IS AN EMR?

The term is frequently used interchangeably with the EHR (Electronic Health Record) and EPR (Electronic Patient Record). An EMR system consists of a database of medical records of several patients stored in digital format. The system can serve patients in one clinic or health center, or they are web-based and accessible by local hospitals, clinics, testing laboratories, and the private practice of physicians. If the EMR is accessible from multiple sites, while maintaining high data security, then the attending physicians can have immediate access to all relevant patient information to make decisions regarding the need for testing, establishing a differential diagnosis, and developing a treatment plan for their patients (Frize, 2012).

8.2   BENEFITS REGARDING THE USE OF EMRS

EMRs enable physicians to access a patient's data before and during the consultation; they can also identify trends by comparing past visits with the current one. They can analyze data and check the literature to support decision-making. If laboratories and pharmacies are networked with the EMR system, then physicians can access results of pathology, diagnostic imaging, and laboratory tests, and are able to order medication directly with the pharmacies. If information is shared between all the health care services needed for a patient's care, such as the physicians and nurse practitioners whom the patient consults, the hospital where out-patient and in-patient services are likely to be provided to the patient, the laboratories and medical imaging facilities where tests are carried-out, the pharmacy where the prescriptions are to be filled, then duplication of procedures and tests will be minimal or avoided and all the care providers who interact with the patient will have the same information, which can help to avoid errors. The system can be designed to produce automated reports, and historical patient data can be retrieved and analyzed on a screen at a faster pace than browsing through a paper file stored in a filing cabinet or retrieved from medical records. Electronic records definitely help to reduce the amount of storage space required by hospitals and clinics (Frize, 2012).

Another positive aspect related to the use of EMRs is patient safety. If the EMR system has the capability to check drug interactions, contraindications, and allergies, this can help to reduce errors related to the prescription of medication. The use of EMRs can also improve adherence to clinical practice guidelines and the delivery of preventive health services. An EMR system that is established for a group practice enables to standardize the format of the records, and their storage and retrieval practices; if appropriate back-ups are maintained, this ensures that no data will be lost; such systems facilitate the management of medical data and can potentially reduce errors. Several systems on the market offer a number of clinical decision support tools; some of these can generate alerts and warnings, deal with prescription refills, and with the management of appointments. In addition, some systems offer education programs for patients, encouraging them to take ownership and management of their illness. Parente and McCullough (2009) found that the EMR system helped to reduce two infections per year at an average hospital attributable to medical care (Parente and McCullough, 2009; Frize, 2012).

In their study of primary care providers, McGuire et al. (2012) found that physicians believed that safety culture improved in terms of the communication with patients and the latter's adherence to their physician's recommendations; these care providers also thought that time constraints was impacted positively. However, there was one exception to this list: recognition of stress. The authors of the study concluded that the "implementation of an EMR in a large primary care practice required redesign of many organizational processes, and was associated with improvements in safety culture. Most primary care providers agreed that the EMR allowed them to provide care more safely" (McGuire et al., 2012; Frize, 2012).

8.3 CONCERNS REGARDING THE USE OF EMRS

Some negative aspects include the need to plan downtime to update and service the product. Some unplanned downtime can also occur in the form of corruption or loss of data or from hackers who get into the system to access confidential information on patients. These concerns accompany all computerized systems. The cost of implementation of such systems is responsible for an initial loss of revenue. Such systems require specialized skill and training for persons who will be doing the data management and for physicians and other care givers who will use the system (Frize, 2012).

Many software options exist on the market, which makes the choice of a product difficult; another issue is the lack of standardization, the busy schedule of physicians and their level of computer skills; if some fields that physicians normally use are missing, this would make data entry difficult; another issue arises for clinicians when they are confronted with complicated interfaces. Other concerns regard the copying, pasting, and duplication of text from one record to the other, for the same patient at different times, or from one patient to another. In their study, Hammond et al. (2003) found that 9% of progress notes contained copied or duplicated text; most of this was benign, but some of the copied text introduced misleading errors into the record which were even possibly unethical or unsafe (Hammond et al., 2003; Frize, 2012).

Solutions to the issues listed above include performing a frequent and regular back-up of the data and ensure that a recovery system is integrated into the EMR system to prevent loss of data. Another recommendation, to prevent hackers getting access to the data and files, is to buy a system that performs an encryption of the data. As mentioned previously, the cost of an EMR system will be recuperated after some time if physicians can order tests and prescribe medications quickly, and access patient information instantly; this is bound to generate a more efficient medical practice. For an EMR system to be successfully implemented, it is critical to ensure that all staff and physicians attend training sessions, to become familiar with the EMR functions. After some practice, data entry should take less time than notes written by hand by the physician or the nurse practitioner. Of course, this depends on how user-friendly the system is and should improve with experience over time. It would help if the system had voice recognition for inputting data dynamically and easily for physicians who would prefer this approach (Frize, 2012).

There are several companies marketing EMR products and the design varies from one company to the other. However, some basic functions must be available in any EMR system. Blumenthal recommends the following: EMRs must contain patient demographics, history of health status, medical examinations, diagnostic test results, charts, images of medical tests, values of vital signs (heart rate, blood pressure, temperature, etc.), and active medications, allergies, smoking and drinking habits, etc. Additionally, they can include drug formulary checks, clinical laboratory results, a reminder to patients of recommended care, and identification and provision of patient-specific health education resources; EMRs can also be used to support the patient's transition between care settings and between various caregivers (Blumenthal and Tavenner, 2010).

Mandl et al. (2001) recommended that the EMR contain at least the following: Problem lists, procedures, allergies, medications, immunizations, history of visits, family medical history, test results, physician and nursing notes, referral and discharge summaries, patient and care provider communications, and patient directives. The records must span a lifetime. More sophisticated systems also offer decision-support tools (Mandl et al., 2001).

8.4   HISTORICAL DEVELOPMENT

In the 1960s, the Mayo Clinic began to develop an electronic medical record system. One of the first more complete medical record systems was developed by Regenstrief in Indiana in 1972. It had both in-patient and out-patient records. Over the years, improvements were made and the system was updated several times.

The Problem-Oriented Medical Information System (PROMIS) was an early EMR system developed at the University of Vermont in 1976 by Lawrence L. Weed and Jan Schultz. PROMIS was an interactive, touch screen system that allowed users to access a medical record within a large body of medical knowledge. At its peak, the PROMIS system had over 60,000 frames of knowledge and was known for its fast responsiveness, especially for its time. Dr. Weed recognized the vast contributions of Harold Cross, MD in setting up a problem-oriented medical practice. Along with the contributions of John Bjorn, MD and Charles Burger, MD, a problem-oriented medical practice was established as a model that formed the basis of electronic record systems (PROMIS).

Dr. Weed's proposed system organized data according to the patient's problem rather than to the source of the data (x-ray, laboratory, etc.). The interface to the patient record was a list of patient problems under which all the data was organized. A sequence of steps was suggested to standardize the protocol of data collection. Subjective data consisted in gathering the patient's story, while objective data consisted of clinical observations. The next step was to evaluate the observations, and the final step was to determine the treatment plan and the care needed. The system had to be concise, complete, reliable, easy to use, and monitored.

In the 1970s and 1980s, several electronic medical record systems were developed and further refined by various academic and research institutions. For example, the initial Technicon system was hospital-based; the Harvard COSTAR system had records for ambulatory care; the HELP system and Duke's "The Medical Record" are examples of early in-patient care systems. Today, there are more than 200 EMR products on the market.

8.5   RATE OF ADOPTION IN WESTERN COUNTRIES

Small physician practices represent the primary source of health care delivery in the U.S. However, in their level of health information technology adoption, these practices lag behind Asia and Europe; they also lag behind integrated medical center and community hospitals in the U.S. A 2011 study reported that, in all the surveys they examined, the majority of physicians claimed that the main barrier to the adoption of such technologies was financial (Police et al., 2011).

A recent survey of primary care physicians in ten countries compared the rate of adoption of EMRs between 2009 and 2012 (Schoen et al., 2012). The survey consisted of interviews which included physicians in general practice and those in family practice in Australia, Canada, France, Germany, Netherlands, New Zealand, Norway, Switzerland, the U.K., and the U.S. As seen in Table 8.1, the rate of adoption varies considerably between countries. This leads to the importance of identifying barriers to the implementation of EMR systems.

Table 8.1: Rate of adoption by physicians in 10 countries in 2009 and 2012 (Source: Schoen et al., 2012)		
Country (sample size)	2009 %	2012 %
Australia (500)	95	92
Canada (2124)	37	55
France (501)	68	67
Germany (909)	72	82
Netherlands (522)	99	98
New Zealand (500)	97	97
Norway (869)	97	98
Switzerland (1025)	--	41
U.K. (500)	96	97
U.S. (1012)	46	69

8.6   BARRIERS TO ADOPTION OF THE EMR

Des Roches and Miralles (2008) identified barriers that prevent a wider adoption of EMRs: 66% of physicians mentioned the initial capital cost; 54% feared the system acquired would not meet their needs; 50% were uncertain of a return on their investment; and 44% feared a quick obsolescence (Des Roches and Miralles, 2008).

Boonstra and Broekhuis (2010) did a systematic literature review of research papers between 1998 and 2009 to identify barriers perceived by physicians that prevented the adoption of EMRs in their practice. The authors categorized the comments into 31 sub-categories that were part of 8 main barriers described below.

8.6.1   FINANCIAL

The most frequently mentioned barrier was financial: physicians wondered if the cost of implementing and running an EMR system was affordable and whether they could gain a financial benefit from it; this aspect was broken into four sub-categories:

1. purchase of hardware and software, and installation; costs ranged between $16,000 and $36,000 U.S. per physician, including $10,000 for the EMR software;

2. system administration, control, maintenance, and support to keep the system working effectively and efficiently which included monitoring, upgrading and maintaining EMRs, and the after-sales service;

3. uncertainty of the return on investment; and

4. financial resources for the initial investment.

8.6.2   TECHNICAL

EMR systems are complex and need computer skills from physicians and providers; there also remain some technical problems with these systems, which lead to complaints by physicians. This creates a resistance to adopt EMRs; some physicians complain of a lack of training and a lack of after-sales support. Physicians also fear that, as time passes, the system will become obsolete and will no longer be useful. They also complain about the lack of customizability and are concerned about loss of access with power failures, computer crashes, or virus attacks. The lack of standardization of software may make it difficult to connect to laboratories, medical imaging departments, and to pharmacies for e-prescribing. Physicians may also wonder if they will be able to connect to other physicians for referrals and get the results back from them. Finally, EMR systems need infrastructure investments such as an internet connection, telephone lines, computer hardware and software.

8.6.3   TIME

This point refers to the time needed to select, purchase, and implement the system, the time to learn how to use it, and the time to enter data remains as a perceived barrier. Physicians also fear they will spend more time with each patient visit; they may be slow in typing notes into the system compared to writing them on paper. The time to convert archived paper records to electronic ones is also a barrier. Several physicians said they were only comfortable with summaries they themselves make from handwritten notes, histories, etc. This may imply that they mistrust the accuracy of a transfer from paper records to an electronic format by someone else.

8.6.4   PSYCHOLOGICAL

More than half of the physicians who do not use EMRs doubted that they could improve patient care and clinical outcomes or the quality of medical practice by using them. Physicians were concerned about the loss of their control over patient information and working processes since these data are shared with and accessed by others; physicians' perceptions of the threat to their professional autonomy was very important in their reaction to EMR adoption.

8.6.5   SOCIAL

Physicians feared that vendors may not be qualified to provide the service needed; vendors may also go out of business. The lack of policies of insurance companies regarding the use of EMRs was perceived as a barrier. Some physicians reported that they sometimes stop using EMRs because hunting for menus and buttons disrupt their clinical encounter. The average screen gaze time of physicians increased from 25% to 55% of the consultancy session, inevitably resulting in less eye-contact and less conversation with the patient. Moreover, as some EMRs are accessible by patients, this might distort the clinical encounter with more interference and distraction from the patient. The perception was that EMRs would change the traditional physician-patient relationship. There was a perception that there was also a lack of support from other physicians, nurses, and administrative staff.

8.6.6   LEGAL

Physicians perceived that unauthorized persons may be able to access the private patient information and this may lead to legal problems.

8.6.7   ORGANIZATIONAL

Physicians in larger practices have a higher adoption rate of EMRs than those in small practices, probably because the former have more extensive support and training and the interconnectivity issue may be more easily solved in larger organizations. Affiliation to a hospital also acted as a facilitator to EMR adoption.

8.6.8   CHANGE PROCESS

Problems occurring during the change process were related to a lack of a proper organizational culture to manage the change, a lack of incentives, individual and local resistance, a lack of community level participation, and a lack of leadership (Boonstra and Broekhuis, 2010).

8.6.9   RESISTANCE TO CHANGE

Lin and Roen (2012) discussed the factors that caused resistance to change among physicians with regards to adopting healthcare information technology. The authors included whether a system was beneficial, useful, or problematic; an example of a problematic system would be one that led to an increased workload or felt unfair or threatening to job security. The study focused on four principal factors: perceived threat, perceived inequity, perceived usefulness, and behavior intention. Emotional resistance included anxiety, frustration, fear, and disquiet which are sometimes found in change processes. Resistance comes from an individual's perception of change as a threat, such as a change of the existing work style, habit, or content.

8.6.10 COMMENTS BY PHYSICIANS

The following comments were made by physicians regarding perceived threat: "I fear that I might lose control over the way I work if I use the EMR"; and "I am worried that I might lose control over the way I order patient tests if I use the EMR"; and "I fear I might lose control over the way I access lab results if I use the EMR" (Lin and Roen, 2012).

Comments regarding perceived inequity were: "Overall, using the EMR would cause more disadvantages than advantages."… "Using the EMR would benefit other staff but not for me."… "Using the EMR would benefit the hospital but not for me" (Lin and Roen, 2012).

Regarding perceived usefulness, comments were: "Using EMRs in my job enabled to complete tasks more quickly."… "Using the EMR improved my job performance."… "Using EMRs enhanced my effectiveness on the job" (Lin and Roen, 2012).

As for behavioral intention, physicians said: "I intend to use the EMR."… "I intend to use more features/modules of the EMR."… "I intend to use the EMR for more of my job responsibilities" (Lin and Roen, 2012).

As can be seen from the discussion above, several efforts are needed by EMR developers and by physician practice organizations to reduce barriers. Several barriers are based on perceptions and those can be eliminated by raising the awareness of physicians on the potential benefits of converting from a paper-based practice to an electronic one. There should also be more sharing among physicians on positive experiences and on how to avoid issues that act as barriers to implementation.

8.7 INCENTIVES TO ESTABLISH AN EMR SYSTEM IN PHYSICIAN PRACTICES

8.7.1 THE UNITED STATES

In the U.S., the Obama administration and the U.S. Congress passed the American Recovery and Reinvestment Act of 2009 (ARRA) in February 2009, which supported the development, adoption, and upgrade of Health Information Technologies (HIT) by authorizing new federal investments in accordance with the development of federal standards. The act provided incentives to EMR adoption among physicians and hospitals, and established a formal framework to support the development of a nationwide infrastructure that would enable the electronic use and accurate exchange of health information. In particular, the Health Information Technology for Economic and Clinical Health Act (HITECH) authorized incentive payments through Medicare and Medicaid to hospitals and clinicians for using EMRs securely to improve patient care. Accompanying these incentives were requirements for "meaningful use" of EMR systems (Blumenthal and Tavenner, 2010; DesRoches and Miralles, 2011). In January 2010, the Department of Health and Human Services (DHHS) proposed requirements for what they called "meaningful use" of EMRs. After receiving some 2000 comments, the DHHS released a final regulation that would apply in 2011 and 2012. This regulation defined a set of core objectives and a set of additional important activities which health care providers could choose during this time period. Suggestions included the use of certified EMR technologies to report on the quality of clinical care, and other measures recommended by the secretary of DHSS (Frize, 2012).

Accompanying the incentives were requirements for "meaningful use" of EMR systems; examples of using a certified EMR technology in a meaningful way were: e-prescribing, the electronic exchange of health information to improve the quality of health care such as promoting care coordination, and reporting on clinical quality and other measures selected by the secretary of Health and Human Services (HHS); all applications had to use certified EMR technology. State Medicaid agencies could develop their own definitions of meaningful use, but they had to be approved by the Secretary of HHS. Further, any State definition that differed from the Medicare criteria had to address populations in the State with unique needs, such as children, and be compatible with State or Federal Administration management systems (DesRoches and Miralles, 2011).

8.7.2   CANADA

An increased adoption of EMR in Canada resulted from investments made by the Canada Health Infoway (CHI), in partnership with provinces and territories that had incentive programs for the adoption and implementation of EMR programs. The statement below explains what is Infoway:

> Canada Health Infoway is an independent, not-for-profit organization funded by the federal government. Infoway jointly invests with every province and territory to accelerate the development and adoption of information and communication technology projects in Canada. Fully respecting patient confidentiality, these secure systems provide clinicians and patients with the information they need to better support safe care decisions and manage their own health. Accessing this vital information quickly helps to foster a more modern and sustainable health care system for all Canadians (**Canada Health Infoway**).

Infoway invests in projects that align with the Pan-Canadian EHR (electronic health record) Blueprint developed by Infoway. But Infoway has no mandate to enforce compliance, as Canadian healthcare is generally governed at the provincial and territorial level. Development of a network of effective interoperable electronic health record solutions across Canada, linking clinics, hospitals, pharmacies, and other points of care, is intended to improve access to health care services for Canadians, enhance the quality of care, and make the health care system more efficient (Canada Health Infoway).

Support from the Federal Government through the 2010 budget, matched by provincial and territorial partners, resulted in more than 11,000 clinicians getting access to EMRs in community-based practices. This was in addition to 25,000 health care providers in ambulatory (outpatient) clinics who also obtained access to EMRs through the program.

In Canada, the EMR is defined as a computer-based medical record in a clinician's practice or organization containing patient demographics, medical and drug history, and diagnostic information such as laboratory results and findings from diagnostic imaging. It is often integrated with other software tools that manage activities such as billing and scheduling.

The benefits of EMR identified in the Canadian project are: health care professionals get timely access to the most up-to-date information when making patient care decisions. Traditionally, laboratory test results, X-rays, medication histories, and other pertinent information have often been in different locations resulting in duplication, extra administration tasks, or relying on a patient's memory, ultimately contributing to delays in clinical intervention and treatment (Canada Health Infoway).

8.8   DESIRABLE DESIGN CHARACTERISTICS OF EMRS

Mandl et al. (2001) developed a comprehensive approach to the design of EMRs. The authors begin with two doctrines to guide the development. (i) EMRs should be designed in a way to be able to exchange all the stored data according to public standards and enable connectivity and interoperability of diverse systems; this principle suggests the use of open standards like HL7 (Health Level 7) which is a voluntary standard for electronic exchange of data on patient admissions, registration, discharge, transfer, queries, orders, results, clinical observations, and billing; other relevant standards in health care are DICOM (Digital Imaging and Communications in Medicine), a standard for storing, printing and transmitting information from medical images; it includes a file format definition and a network communications protocol; and X12 (www.x12org) for authorization, referral and billing records; note that the data can then be stored and used with proprietary tools. (ii) Patients should have control over who accesses their EMR file; this is intended to protect their privacy according to their preferences and prevent some of the current misuses of personal medical information; this may increase the trust that patients have for EMR use. Some of the EMR software providers want to own the patient data, but if patients control its creation, annotation, modification, dissemination, use, and deletion, this would ensure patient access to their data while protecting privacy (Mandl et al., 2001).

Mandl et al. (2001) also proposed a number of design characteristics that EMRs should have:

- **Comprehensiveness.** This refers to a list of what EMRs should contain, which was provided in an earlier section. The records should ideally span a lifetime to be available for a retrospective analysis when needed.

- **Accessibility.** EMRs are needed not just for a visit to the physician, but also in cases of emergency, or when patients are referred to a specialist or for some tests (laboratory, imaging, cardiac, etc.). Also, with a patient's permission, de-identified data should be able to be used by researchers and public health authorities.

- **Interoperability.** EMR information should be shared with different computerized medical systems; this means sending and accepting data to and from multiple sources, otherwise the EMR remains fragmented and will not be very useful.

- **Confidentiality.** There is a balance to keep between providing access to the EMR, with the desire to provide informed medical care, and maintaining privacy. There should be an "overriding" clause to provide access to authorized providers in cases of emergency.

- **Accountability.** Mandl et al. (2001) recommended that patients should not be able to delete or alter information entered by others, but they should be able to annotate and challenge interpretations in their record. Patients should also be able to see who accessed a part of the record, under what circumstances and for what purpose. Reliable authentication is essential and consent needs to be obtained if the data is to be used for research purposes (Mandl et al., 2001).

Another consideration is the transfer of records to a new computer system. Computers are frequently replaced with machines that have more advanced capabilities. This must be done carefully with privacy and confidentiality in mind to ensure deletion of old records in a complete and safe manner after the transfer to the new system. Deleting files from a computer does not completely erase them. More needs to be done to the memory and to the back-up systems.

8.9   PATIENT SAFETY

A study found that the EMR decreased the time to develop a synopsis of the patient and improved communication efficiency, which can be said to be related to safety (Joos et al., 2006). A study that surveyed 103 primary care providers concluded that implementation of an EMR system in a large practice required the redesign of many organizational processes, but resulted in improvements in the care givers' ability to provide care more safely (McGuire et al., 2012). These authors argue that EMR systems can help health care organizations and their professionals to organize data and improve workflows.

8.10   ETHICAL CONSIDERATIONS

8.10.1   AUTONOMY

The principle of autonomy infers the duty to respect the individuals' right to make their own decisions. In the case of EMRs, autonomy would mean that patients could have the right to decide whether their records can be kept in that manner and with who the record is shared; ideally, it also means that patients should have access to the health information it contains and it may include the right for patients to add comments or annotations to their file.

It is common for people to manage their bank accounts, investments, purchases, and many other services online; many also use the web as a source of health-related and medical information, so expect to extend this type of control to their own medical information. This would also apply to guardians or parents in cases of persons who are mentally disabled or minors. Patients should have access to all the information they need to make appropriate decisions regarding their illness management; in this way, they can make informed choices when they face treatment alternatives; or be encouraged by their compliance with the physician's instructions. If part of the information is withheld, this removes the right of autonomy from the person making the decision. Ultimately, it would be ideal if there is an agreement between patients and physicians, and a mutual trust on sharing the information with appropriate partners of the care providers such as laboratories, pharmacies, etc. There must also be safeguards such as strict and transparent rules that guide the access to medical records, a mechanism for complaints, and understandable information about the program (Frize, 2012).

8.10.2   BENEFICENCE AND NON-MALEFICENCE

Beneficence refers to the duty to maximize benefits; while non-maleficence refers to the duty to do no harm. The benefits of using EMRs have been discussed in detail earlier in the chapter. Since much data are collected today in electronic format, another benefit would be to remove identifiers from the data so that it can be shared with researchers who could improve health care delivery through analysis of the data. An example of this is provided in Chapter 9 of Part II, on clinical data collection, storage, and retrieval.

Regarding non-maleficence, this concept applies to potential physical, psychological, and spiritual harm. Some ways in which this can apply to EMRs are: patients may be upset after reading the physician's notes; they may be concerned about how the information in their record is shared, and with whom. However, others may be pleased at the efficient way in which their medication is prescribed; another attitude may be that patients will take more responsibility in adhering to the physician's recommendations and treatment plan (Frize, 2012).

8.10.3 JUSTICE

This refers to the duty to treat everyone fairly, and to distribute risks and benefits as equally as possible to all persons. As seen in previous sections, EMRs have the potential to provide a safer and higher quality of care to a large number of people, which would make the system seem equitable. But since not all persons have access to internet and computers, there could be some injustice if the EMR system was meant to provide access to patients. When patient access becomes part of the EMR system used by the care providers, then efforts will be needed to ensure that as many people as possible gain access to their records (Frize, 2012).

The next chapter discusses the knowledge management (KM) process with application to the health care domain. It covers the first step in this process which consists of data acquisition, storage, and retrieval.

REFERENCES

Blumenthal, D, Tavenner, M (2010). "The 'Meaningful Use' Regulation for Electronic Health Records." *NEJM*, 363:501-504. Available at: http://www.nejm.org/doi/full/10.1056/NEJMp1006114; last accessed July 2013. DOI: 10.1056/NEJMp1006114.

Boonstra, A, Broekhuis, M (2010). "Barriers to the Acceptance of Electronic Medical Records by Physicians from Systematic Review to Taxonomy and Interventions." *BMC Health Services Res.*, 10:231. Available at: http://www.biomedcentral.com/1472-6963/10/231; last accessed July 2013. DOI: 10.1186/1472-6963-10-231.

Canada Health Infoway (2013). Available at: https://www.infoway-inforoute.ca/index.php/progress-in-canada/benefits-realization#emr_benefits; last visited June 2013.

DesRoches, CM, Campbell, EG, Rao, SR, Donelan, K, Ferris, TG, et al. (2008). "Electronic Health Records in Ambulatory Care — A National Survey of Physicians." *NE JM*, 359:50-60. Available at: http://www.nejm.org/doi/full/10.1056/NEJMsa0802005#t=article; last accessed July 2013. DOI: 10.1056/NEJMsa0802005.

DesRoches, CM, Miralles, PD,(2011). "Meaningful Use of Health Information Technology: What Does it Mean for Practicing Physicians?" Chapter 1: 1-14. In Skolnik, NS. Ed. *EMRs : A Practical Guide for Primary Care.* Humana Press, Springer.

Frize, M (2012). "Electronic Medical Records (EMRs): Patient Safety and Ethical Considerations." *Eth. Biol. Eng. Med.*, 3(1-3): 3-8. DOI: 10.1615/EthicsBiologyEngMed.2013006933.

Hammond, KW, Helbig, ST, Benson, CC, Brathwaite-Sketoe, BM (2003). "Are Electronic Medical Records Trustworthy? Observations on Copying, Pasting and Duplication." *AMIA Ann. Symp. Proc.*: 269-273.

Joos D, Chen Q, Johnson, KB (2006). "An Electronic Medical Record in Primary Care: Impact on Satisfaction, Work Efficiency and Clinic Processes." *AMIA Annu. Symp.*: 394-398.

Lin, C, Lin, IC, Roan, J (2012). "Barriers to Physicians' Adoption of Healthcare Information Technology: An Empirical Study on Multiple Hospitals." *J. Med. Syst*, 36:1965–1977. DOI 10.1007/s10916-011-9656-7. DOI: 10.1007/s10916-011-9656-7.

Mandl, KD, Szolovits, P, Kohane, IS (2001). "Public Standards and Patients' Control: How to Keep Electronic Medical Records Accessible but Private." *BMJ*, 322: 283. Available at: http://www.bmj.com/content/322/7281/283.full; last accessed July 2013. DOI: 10.1136/bmj.322.7281.283.

McGuire, MJ, Noronha, G, Samal, L, Yeh, HC, Crocetti, S, Kravet, S (2012). "Patient Safety Perceptions of Primary Care Providers after Implementation of an Electronic Medical Record System." *J. Gen. Intern. Med.* Available at: http://link.springer.com/article/10.1007%2Fs11606-012-2153-y?LI=true; last visited Dec 2012) DOI: 10.1007/s11606-012-2153-y.

Parente, ST, McCullough, JS (2009). "Health Information Technology and Patient Safety Evidence From Panel Data." *Health Aff.*, March/April; 28(2): 357-360. DOI: 10.1377/hlthaff.28.2.357.

Police, RL, Foster, T, Wong, KS (2011). "Adoption and Use of Health Information Technology in Physician Practice Organizations: A Systematic Review." *Informatics Prim. Care*, 18: 245–58.

Schoen, C, Osborn, R, Squires, D, Doty, M, Rasmussen, P, Pierson, R, Applebaum, S (2012). "A Survey of Primary Care Doctors in Ten Countries Shows Progress in Use of Health Information Technology, Less in Other Areas." *Health Aff.*, 31(12): 1-12. Available at: http://content.healthaffairs.org/content/early/2012/11/13/hlthaff.2012.0884.full.html; last accessed July 2013. DOI: 10.1377/hlthaff.2012.0884.

OTHER SUGGESTED READING

Häyrinen, K, Saranto, K, Nykänen, P (2008). Definition, Structure, Content, Use and Impacts of Electronic Health Records: A Review of the Research Literature. *Intl. J. Med. Inform.*, 7 7: 291–304. DOI: 10.1016/j.ijmedinf.2007.09.001.

Skolnick, NS, Series editor. (2011). *Electronic Medical Records (EMRs): A Practical Guide for Primary Care.* Humana Press (Springer).

Wang, SJ, Middleton, B, Prosser, LA, Bardon, C, Spurr, CD, Cardichi, PJ, et al. (2003). "A Cost-Benefit Analysis of Electronic Medical Records in Primary Care." *Amer. J. Med.*, 114(5): 397-403. DOI: 10.1016/S0002-9343(03)00057-3.

Willison, DJ, Keshavjee, K, Nair, K, Goldsmith, C, Holbrook, AM. "Patients' Consent Preferences for Research Uses of Onformation in Electronic Medical Records: Interview and Survey Data." *BMJ*, 15 February: 326:373. DOI: 10.1136/bmj.326.7385.373.

Van Wingerde, FJ, Schindler, J, Kilbridge, P, Szolovits, P, Safran, C, Rind, D, et al., (1999). "Legal Issues Concerning Electronic Health Information: Privacy, Quality, and Liability." *JAMA*, 282:1466–1471. DOI: 10.1001/jama.282.15.1466.

Winkelman, WJ, Leonard, KJ, Rossos, PG, (2005). "Patient-Perceived Usefulness of Online Electronic Medical Records: Employing Grounded Theory in the Development of Information and Communication Technologies for Use by Patients Living with Chronic Illness." *JAMIA*, 12:306-314. Available at: http://jamia.bmj.com/content/12/3/306.full; last accessed July 2013. DOI: 10.1197/jamia.M1712.

CHAPTER 9

Knowledge Management (KM) in a Clinical Environment: Data Acquisition, Storage, and Retrieval

This chapter briefly describes the four key steps in the knowledge management (KM) process and provides details on the first step which concerns the acquisition, storage, and retrieval of quality clinical data for future analysis and processing. The second step is knowledge discovery (KD), discussed in Chapter 10; it describes a number of artificial intelligence tools and analytical approaches. The third step is knowledge translation (KT), that is, how knowledge is presented to the users. This includes the development and use of clinical decision support systems (CDSSs). The fourth step is knowledge integration and sharing (KIS); for this Web Services and the Semantic Web Services are presented as a means to integrate information and share it with users from any of the practical locations where they may need the information; these last two steps are discussed in Chapter 11. Throughout this book, the example selected to illustrate the four steps of knowledge management is the perinatal care clinical environment, more specifically related to obstetrics and preterm infants hospitalized in a neonatal intensive care unit (NICU).

9.1   THE IMPORTANCE OF INTEGRATING KNOWLEDGE MANAGEMENT (KM) INTO CLINICAL CARE

The quick tempo of illness in critically ill patients has spawned numerous monitoring technologies which generate large volumes of information. However, these monitors and hospital information systems are often unconnected, uncoordinated, and can independently generate multiple alarms. This highlights the importance of developing techniques to integrate the large volume of data into health care knowledge. The section below explains how physicians arrive at a diagnosis and a patient management plan.

When a patient needs medical attention, the physician interacts with the patient through the construct of the medical model, which follows a structured interview and examination technique to establish a diagnosis and develop a treatment plan. Before establishing a definitive diagnosis and treatment plan, the physician starts with what is called a differential diagnosis, an investigative plan, and a potential treatment plan. During the phase of the differential diagnosis, the physician lists likely pathologies which may explain the patient's condition. The investigative plan deals with the uncertainty present in the differential diagnosis; thus at this stage, the physician recommends a number of tests in order to "rule out" or "rule in" the conditions listed in the differential diagnosis.

With the treatment plan, an attempt is made, using a variety of interventions, to return the pathologic state to as normal a physiologic state as possible. The modern interaction between the physician and the patient is generally supplemented by tests of body fluids, medical imaging, and electrical signal recording and analysis. The results of the tests enable the physician to determine more precisely the likelihood of the presence of a disease and to monitor the effectiveness of the treatment. If the patient is healthy or the diagnosis and prognosis are known, then no testing is needed. Even if a definitive diagnosis is established, how the illness unfolds over time is often uncertain and may be different from a patient to another. The tests help to deal with the uncertainty in the diagnosis and management of patients (Frize et al., 1995; Frize et al. 2007).

9.2 THE HEALTH CARE KNOWLEDGE MANAGEMENT (KM) PROCESS

The four-step knowledge management process described in this book link the acquisition of data as clinical databases, data mining techniques to analyze the data, clinical decision support tools, and technologies that integrate and share the information between caregivers. Examples to illustrate the process are taken from the perinatal care environment.

9.2.1 STEP 1: ACCESS TO CLINICAL DATA: DATA COLLECTION, STORAGE, AND RETRIEVAL

The first step in the KM process is the collection, storage, and retrieval of data. Today, with the ubiquitous presence of electronic medical records (EMRs), hospital information systems (HIS), Picture Archiving and Communication Systems (PACS) used in many medical imaging departments, and numerous monitoring devices that generate large volumes of information, it has become more common to collect all of these data, signals, and images into clinical databases. However, in order to be able to perform useful and valid analysis, it is critical that the data collected be of the highest quality, especially when these are integrated into clinical decision-support tools. To ensure data quality and integrity, regular data checks need to be done (also called data audits).

Developers of clinical decision support tools need to have access to clinical databases, therefore they need to have a strong and trusting partnership with caregivers. This can be quite challenging with the extensive privacy regulations now in place in Canada and the U.S. to ensure the confidentiality of patient information. Prior to obtaining clinical data, the partners (physicians and artificial intelligence specialists) must apply to the appropriate ethics review boards in all institutions concerned (hospital, university, etc.), and obtain clearance for the work proposed. Involving the clinical personnel at all stages of the research and development work is essential for its ultimate success; experience has shown that health care practitioners are more likely to incorporate information technology (IT) tools into daily practice if they are involved in the conception and development of the tools at each and every step (Frize et al., 2007; Frize and Weyand, 2010).

9.2.2 STEP 2: KNOWLEDGE DISCOVERY (KD)

In this step of the KM process, a variety of data mining tools are used to analyse the data, or to extract features, observe trends, develop prediction models for assessing potential clinical outcomes or changes in the health status of the patient; some of these systems can issue warnings and/or alarms based on a set of criteria that can vary from patient to patient and by the type of signal assessed. The KD step transforms data into knowledge. Data mining tools that have shown promising results are: Statistical techniques, especially logistic regression; artificial neural networks, case-based reasoning, fuzzy-logic sets, and genetic algorithms. A more detailed discussion of some of these tools in the KD process is found in Chapter 10.

9.2.3 STEP 3: KNOWLEDGE TRANSLATION (KT)

In this step, the output of the KD tool is transformed into information that could be useful for clinicians. One approach to achieve this goal is to design graphical user interfaces (GUI) that can present the information in a way that is intuitive and user-friendly. Examples of useful features that could be included in KD tools are: Providing a probability attached to predictions made in order to enable clinicians to assess the level of trust they can assign to the tool's output; initiating alerts for caregivers when the estimated outcomes present a clinical situation that urgently needs attention or where a meaningful change is measured; warnings could also be provided for a clinical situation that is deteriorating without being critical. KT can include the development of clinical decision-support systems (CDSSs), which are becoming increasingly part of the information technology applications in the delivery of health care. More details are provided in Chapter 11 on the KT step of the KM process.

9.2.4 STEP 4: KNOWLEDGE INTEGRATION AND SHARING (KIS)

Results from the KT process should be shared with all practitioners involved in patient care in the particular environment where the tools are to be implemented and used. A promising technology to achieve this integration and sharing of information is the Web services (intranet or internet) to provide caregivers easy and quick access to data, analyses, decision-support tools, and clinical alerts. Access should be provided at the point of care and in remote locations, such as a caregiver's office or home; this would enable clinicians to utilize any of the steps of the KM process wherever they are (Frize et al., 2007).

An important aspect to ensure success of any of the steps and tools used in the KM process is to position them to work within the traditional medical model presented at the beginning of this chapter. A few examples follow: If a Case-Based Reasoning (CBR) tool can find the ten closest-matching patient cases to a newly admitted patient, this can simulate a physician thinking: "I have seen patients like this…and this is what happened to them." Similarly, an artificial neural network that predicts a clinical outcome, for example whether the patient is expected to live or die, simulates a physician thinking: "And for this patient, this is what I think will happen."

It is important that clinical decision-support tool developers understand that they must never attempt to replace physicians and caregivers by their systems. For a successful implementation of information technology tools in a real clinical environment, they need to remain as close as possible to the way physicians perform their diagnosis and patient management, fit well into their workflow, and not add much data entry or keyboard tasks. Information technologies applied to KM should be intuitive, easy to use, and provide information quickly and effectively (Frize et al., 1995; Taylor et al., 1993; Frize and Weyand, 2010). More details are provided in Chapter 11 on the KIS step of the KM process.

9.3   CLINICAL DATA REPOSITORY DESIGN

Considering the multiple sources of data in a clinical environment, it is important to integrate these into one repository which will be called Clinical Data Repository (CDR) in this chapter. However, the complexity and fast evolution of clinical information make development and maintenance of medical databases challenging. Therefore the design of a CDR should be flexible enough to handle new data types and attributes without having to change the basic design. More details are provided on a CDR design below.

A CDR enables to collect data from different clinical systems, such as laboratory, pharmacy, medical imaging, pathology, admissions and discharge information, diagnoses, and discharge summaries. CDRs can also collect information from cardiac monitors, ventilators, oximeters, and other devices connected to patients that have an output compatible with standards in the field.

It is a certainty that CDRs will become more common in clinical environments as more and more medical devices enable to collect data in electronic form; a CDR can help to integrate data from various sources (Mullins et al., 2006). Having relevant patient data aggregated into a single location like a CDR facilitates the implementation of clinical decision support tools.

Some clinical data repositories are provided by manufacturers with their equipment, especially when monitors are connected to a central station; but these systems are usually proprietary and thus can only be accessed via the client software built specifically for a specific database. Due to the proprietary nature of these systems, the data cannot be directly accessed for research applications and analysis; moreover, users are limited to applications provided by the equipment manufacturer and no new functionality can be easily added. Moreover, data gathered for clinical use is typically only stored for a few days, which is generally fine for clinicians to observe trends in the variables they are monitoring, but definitely not appropriate for clinical research (Gilchrist et al., 2008).

Conventional databases typically have a single row for each entity (e.g., a patient), a column for each attribute associated with the entity (e.g. gender, birth date) and the value is stored in the table cell where the attribute column and the entity row intersect. When new data types or attributes are added, the physical design of conventional databases must be changed; new columns need to be added to store the new attribute values. This can be time consuming for the maintenance and upkeep of the database. A CDR should be flexible enough to handle new data types and attributes without changing the physical database schema. The structure of conventional databases does not make it easy to store time-varying data, or features with different numbers of measurements (e.g., minute-to-minute heart rate compared to blood gases taken only a few times per day). An example of a CDR using open-source tools to collect data in real-time and store them for many years is described below; this approach can create large data sets for use in both research and clinical work (Gilchrist et al., 2010; Gilchrist et al., 2011; Frize et al., 2013).

Databases using the Entity-Attribute-Value (EAV) format provide a flexible and simple schema to store biomedical data. All data can be stored in one generic table with three columns: entity, attribute, and value. EAV is often used when a large amount of heterogeneous data must be represented consistently, and when the number of attributes and types do not need to be pre-determined. Medical data frequently contain mixed or combined data types with digits, characters, images, and signals. To incorporate the time at which the value was measured, a fourth column was added (Entity-Attribute-Value-Time = EAVT). The performance of the hybrid CDR was tested against two other EAV designs: the simple EAV and the multi-data type. The hybrid design allows the use of multiple data types in a single table by using multiple value columns in a single table. By using NULL values in the columns that are not used, less space is wasted as a NULL value requires very little memory. The hybrid EAVT format uses integer (4 bytes), floating point (4 bytes), and text (2 to N bytes) values in a single table. The three designs were compared in terms of storage capacity, query speed, and query complexity that could be handled (Gilchrist et al., 2011; Frize et al., 2013).

Figure 9.1 shows data sent to the CDR from the patient monitors, the Admissions-Discharge-Transfer (ADT) system, and the laboratory systems, using the HL7[1] messaging protocol. MIRTH,[2] and open-source HL7 interface engine are used to filter the data, and pull out relevant information to be stored in the CDR database. Confidential patient information is automatically segregated from the research data using a study ID to associate the patient with the research data in an anonymous way. This automatic segregation allows researchers to access only the de-identified medical data without requiring any further processing or scrubbing of the data. The performance of the three EAV formats was evaluated using queries of millions of clinical data points of real patients

---

[1] Hl7: Health Level Seven International; online at www.hl7.org
[2] Mirth Corporation; online at: www.mirthcorp.com/

captured from patient monitors and other systems in the neonatal intensive care unit (NICU) at the Children's Hospital of Eastern Ontario (CHEO) (Gilchrist et al., 2011).

Adding a new row of data is simple: since the design of the database does not change, new data can fit into existing columns. However, to add a new attribute to a conventional database, such as blood pressure, the physical schema needs to be modified since a new column for the attribute is needed. This changes every row that already exists in the database, and can affect applications that use the database (Gilchrist et al., 2011).

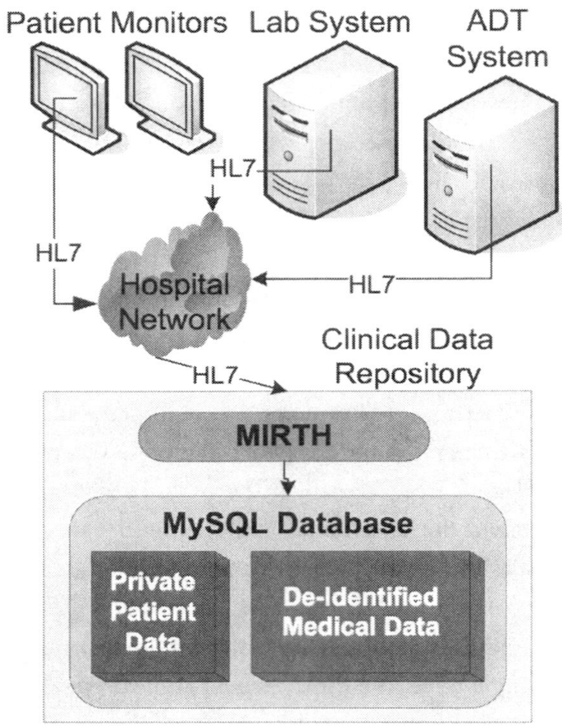

Figure 9.1: Data collection and storage. (Source: Gilchrist et al., 2011; Gilchrist, 2012)

In terms of storage performance, the hybrid format made the most efficient use of space; it uses a single table and can store data in their original format; the simple format also uses a single table, but it stores everything as text, which is less compact for storing integers or floating point values. In terms of query speed, the time taken for querying 282,584 data points from one patient was 2.87 s for the simple EAV; it was 2.81 s for the hybrid EAV format, and 8.44 s for the multi-data type format, three times slower for this type of query. The multi-data type format uses separate tables for each data type that need to be combined using union operators when querying all data from a patient. These extra operations are not needed for the simple or hybrid formats. An analysis showed that the hybrid EAV format allows the use of multiple data types in their original formats, while requiring the least amount of storage space, and simplifies queries compared to the multi-data type format; the performance is essentially the same as the simple EAV type and faster than the multi-data type format (Gilchrist et al., 2011).

For the multi-data type, when a large number of attributes needs to be stored in a conventional database, the attributes are usually split into multiple tables. Accessing data in multiple tables requires the use of the JOIN function which can reduce querying efficiency considerably (Anhoj, 2003). The empty cells in a conventional database can result in wasted space if the database software does not support using NULL values. If the data are very sparse, a considerable amount of wasted space may result (Gilchrist et al., 2011; Gilchrist, 2012; Frize et al., 2013).

9.4 EXAMPLE OF PERINATAL DATABASES USED IN A RESEARCH PROJECT

This section describes three databases, two of which consist of data on newborns and on their mothers during the pregnancy, and the other consists of a database of patients in a neonatal intensive care unit.

9.4.1 THE PERINATAL PARTNERSHIP PROGRAM OF EASTERN AND SOUTHEASTERN ONTARIO (PPPESO) DATABASE

Until 2009, the PPPESO (Perinatal Partnership Program of Eastern and Southeastern Ontario) collected the Niday database. This program has now been replaced by the BORN program (Better Outcomes Registry and Network). The Niday database was used to predict pre-term birth, delivery type (Caesarian section or vaginal), Apgar score of the infant (at 5 min after birth), and mortality. The PPPESO worked together with hospitals, health departments, community agencies, academic institutions, private practitioners, and consumers to effectively link perinatal care, education, and research to identify issues, develop and implement solutions that improve evidence-based regionalized perinatal care to childbearing families in the region. In 2001, 17,406 women gave birth in hospitals in this region. The list of variables collected included birth date, mother's age, postal code, inter-hospital transfer, number of previous term/preterm babies, number of babies, baby's sex, baby's weight, presentation, monitoring methods used, maternal pain relief offered, antenatal steroids used, whether scalp or cord blood gases were performed, maternal intention to breastfeed, maternal smoking, and neonatal transfer. Outputs of interest included: number of weeks gestation, labor type, Apgar score at 1 and 5 min, newborn resuscitation techniques, and neonatal death/stillbirth.

9.4.2 PRAMS (PREGNANCY RISK ASSESSMENT MONITORING SYSTEM) DATABASE

A surveillance project was initiated by the Center for Disease Control (CDC in Atlanta, U.S.), to collect state-specific, population-based data on maternal attitudes and experiences before, during, and shortly after pregnancy. On a yearly state-wide basis, PRAMS selects a sample of all women who gave birth that year, creating a heterogeneous subset of mothers, so that the findings can be extrapolated to an entire state's population. Using a standardized data collection methodology, PRAMS (Centers for Disease Control and Prevention) combines two modes of data collection: a survey conducted by mailed questionnaire with multiple follow-up attempts, and a survey by telephone. Along with obstetrical and maternal medical history, PRAMS provides other information such as: barriers to prenatal care, maternal use of alcohol and cigarettes, physical abuse, contraception, economic status, sources of maternal stress, social support and services, and mental health during pregnancy (CDC, PRAMS).

9.4.3 THE CANADIAN NEONATAL NETWORK (CNN) DATABASE

The CNN provided data collected across Canada from 1996–1997 (17 centers, about 20,000 infants and 1,000 deaths.) The CNN database contains several subsets: admission data; obstetric characteristics; illness severity information; discharge data elements; diagnoses and conditions; and disposition. It includes illness severity scores; data elements abstracted prospectively while the patient is in the NICU; diagnoses; complications; procedures and therapies; and a discharge abstract completed after death or discharge of the patient from the last hospital to which the patient has been admitted. Outcome variables include: in-hospital death; major morbidities; resource utilization. The current studies are: prediction of mortality in neonates of various gestations, duration of assisted ventilation and length of stay (LOS), and complications such as: bronchopulmonary dysplasia, intraventricular hemorrhage, necrotising enterocolitis, and retinopathy of prematurity.

The CNN variables are all recorded in raw numeric form rather than categorical form for ease of abstracting and analytic flexibility. The data are entered in a standardized format to study illness severity, practice variations, and resource consumption. Uniform conditions are established for all data collection by a comprehensive manual. The data are collected by a Research Assistant who visits NICUs daily and enters data from the patient record and other unit records directly into a laptop computer; the program permits error checking. Case numbers are assigned sequentially and can be cross-referenced with a code assigned by the local centre for each patient. The system avoids paper and keyboard entry, duplicate assignment of study number, and filing/storage of data forms, while improving security and reliability. Obstetric information is verified from maternal records. The discharge abstract is completed after death or discharge of the patient from the last hospital to which the patient has been admitted, as many patients are retro-transferred to other hospitals. Summary data, collected every two weeks, include staff-patient ratios and nursing acuity scores for all patients. Data collected annually include nursing wage rates, bed occupancy rates, patient turnover rates, adoption of new technologies in each unit (including certain ventilator technologies, pulmonary function monitoring, and magnetic resonance imaging), number of assisted reproduction births, total deliveries in the area, and NICU admission rate/1,000 deliveries. All data are cleaned on receipt at the co-ordinating center by removing out-of-range entries. The CNN study includes re-abstraction of a random 5% of charts at each participating institution to check reliability.

These high quality databases have enabled to develop several models that estimate clinical outcomes and predict resource utilization in perinatal medicine environments. The next chapter discusses the second step of the KM process—knowledge discovery—that includes data mining and data analysis. The examples used are from the perinatal environment.

REFERENCES

Anhoj, J (2003). "Generic Design of Web-Based Clinical Databases." *J. Med. Internet Research*, 5(e27), November. DOI: 10.2196/jmir.5.4.e27.

Catley, C (2007). "An Integrated Hybrid Data Mining System for Preterm Birth Risk Assessment Based on a 'Semantic Web Services for Healthcare' Framework." Doctoral Thesis, Systems and Computer Engineering, Carleton University, Ottawa, Canada K1S 5B6.

CDC, PRAMS (n.d.). "Centre for Disease Control, Atlanta." Available at: http://www.cdc.gov/prams/

CNN (n.d.). "Canadian Neonatal Network." Available at: http://www.canadianneonatalnetwork.org/portal/; last accessed July 2013.

Frize, M, Solven, FG, Stevenson, M, Nickerson, BG, Buskard, T, Taylor, KB (1995). "Computer-Assisted Decision Support Systems for Patient Management in an Intensive Care Unit." *Proc. Medinfo'95*, Vancouver: 1009-1012.

Frize, M, Weyand, SA, Bariciak, E (2010). "Suggested Criteria for Successful Deployment of a Clinical Decision Support System (CDSS)." *Proc. Med. Measure. Appl.*, Ottawa: 69-72. DOI: 10.1109/MEMEA.2010.5480227.

Frize, M, Bariciak, E, Gilchrist, J (2013). "PPADS: Physician-PArent Decision-Support for Neonatal Intensive Care." *Proc. Medinfo 2013*: 23-27.

Gilchrist, J, Frize, M, Bariciak, E, Townsend, D,(2008). "Integration of New Technology in a Legacy System for Collecting Medical Data - Challenges and Lessons Learned." *Proc. 30th Intern. IEEE Eng. Med. & Biol. (EMBC)*, Vancouver: 4326-4329. DOI: 10.1109/IEMBS.2008.4650167.

Gilchrist, J, Frize, M, Ennett, CM, Bariciak, E (2010). "Performance Evaluation of Patient Monitor Data Queries Comparing EAV Storage Formats in a CDR". MeMeA (Medical Measurements and Applications Workshop). Ottawa, May:63-68.

Gilchrist, J, Frize, M, Ennett, CM, Bariciak, E (2011). "Performance Evaluation of Various Storage Formats for Clinical Data Repositories." *IEEE Trans. Instrum. Measure.*, 60 (10): 3244-3252. DOI: 10.1109/TIM.2011.2122850.

Gilchrist, J (2012). "A Clinical Decision Support System using Real-Time Data Analysis for a Neonatal Intensive Care Unit." PhD Thesis, Systems and Computer Engineering Department, Carleton University, Ottawa, ON, Canada.

Mullins, I. M., M. S. Siadaty, J. Lyman, K. Scully, C. T. Garrett, W. G. Miller, R. Muller, B. Robson, C. Apte, S. Weiss, I. Rigoutsos, D. Platt, S. Cohen, and W. A. Knaus (2006, December). Data mining and clinical data repositories: In- sights from a 667,000 patient data set. Computers in Biology and Medicine 36, 1351–1377. DOI: 10.1016/j.compbiomed.2005.08.003.

Taylor, KB, Nickerson, BG, Frize, M, Solven, FG, Dunfield, V (1993). "Use of Case-Based Reasoning to Assist Patient Management in an Intensive Care Unit." *CMBEC/CCGB Conference*, Ottawa: 248-249.

OTHER SUGGESTED READING

Bali, RK, Dwivedi, AN (2007) *Health Knowledge Management. Issues, Advances, and Successes.* Springer. Health Informatics Series.

CHAPTER   10

Knowledge Discovery (KD): Data Analysis and Data Mining Tools

This chapter focusses on the second step of the knowledge management (KM) process: knowledge discovery (KD). Various approaches exist for analyzing medical data. The chapter presents some of the severity of illness scores used in predicting clinical outcomes, followed by a short description of data mining tools developed to predict outcomes.

10.1   SCORING SYSTEMS TO ESTIMATE OUTCOMES

In past decades, models that predict outcomes or assess the status of a patient initially used scores based on a combination of demographic, therapeutic and physiologic variables. One example in neonatal care is the Score for Neonatal Acute Physiology (SNAP), based on the adult intensive care unit prediction model "APACHE" (Acute Physiology and Chronic Health Evaluation (Knaus, 1981; Richardson et al., 1993). The SNAP Score weights range from 0–5 and have 37 possible measurements chosen through clinical expertise. SNAP-II was developed by logistic regression analysis using the SNAP Score variables and data from the seventeen NICUs comprising the membership of the Canadian Neonatal Network (CNN) described in Chapter 9. SNAP-II achieves a similar performance to SNAP while only requiring six of the 37 variables (lowest blood pressure, lowest serum, pH, lowest temperature, lowest pO2/FiO2 ratio, seizures, and urine output—all variables collected in the first 12 h from admission to the NICU) (Richardson et al., 2001). These scoring systems classify patients according to their severity of illness, which correlates with mortality. Death is a rare outcome in the NICU given that the overall mortality rate in the CNN database is less than 4% (Lee et al., 2000).

Although illness severity and therapeutic intensity scores have been used as research tools in the past, they have important limitations. For example, at its highest score, the predictive power of SNAP is about 50% (Meadow and Lantos, 2003). Scoring systems are not easily updated to reflect new practices, nor are they designed to be specific to the practice in a particular unit or region. None have been shown to be sufficiently accurate in predicting the outcome of individual patients to support clinical or ethical decision-making. Moreover, all scores have limitations in the outcomes they can predict and they do not appear to have any utility in improving diagnostic accuracy or in reducing medical errors. Richardson and Tarnow-Mordi (1994) reviewed 30 neonatal scoring systems and found that few had been validated on large, concurrent samples of newborns. Scoring systems have been compared to help choose such tools, but it was reported that many clinicians remain skeptical about using scoring models in actual patient care (Hyzg, 1995). Indeed, scoring systems appear to be little better than clinical judgment in predicting the probability of death, with a probability approaching 50%. An automatic system to predict important outcomes with adequate sensitivity and specificity, based on a unit or a region-specific database, may have greater utility as a decision-support tool in the NICU (Frize et al., 2007). To replace scoring systems, several data mining and analytical tools have been developed and are described in the sections below. Accurate outcome predictions for single patients in the NICU would facilitate clinical and ethical decision-making, patient management, provision of accurate parental information, and appropriate resource allocation planning.

10.2 EXAMPLES OF KNOWLEDGE DISCOVERY TOOLS

Research efforts have focused on the application of information technology and artificial intelligence, including cased- based reasoning (CBR), artificial neural networks (ANN), decision trees (DT), and knowledge-based systems (KBS) to provide easy access to information and knowledge management at the point of care. The literature on the use of ANNs to predict outcomes for a variety of disease conditions in adults is numerous; however, there are few studies using ANNs in prediction of outcomes for preterm newborns. The ANN approach has been reported to outperform the statistical method by a small margin, suggesting that these approaches may be complementary to each other (Ambalavanan and Carlo, 2001; Sargent, 2001). A study by Lisboa (2002) discussed the health benefits of using ANNs in medical intervention. Lisboa's survey included randomized and non-randomized clinical trials of ANNs in the domains of oncology, critical care, cardiovascular medicine, and included four studies in perinatal or neonatal care. The review noted the potential for extensive benefit but criticized the poor methodology and the exaggerated claims made in some of the studies (Lisboa, 2002).

The design blueprint for decision-support research contained in Lisboa's review had the following components: clarify the purpose of the study; model design (particularly to control for over-fitting); network regularization (e.g., using a Bayesian regularisation framework); variable selection (number of observations should be 5-10 times the number of available covariates); validation: support for learned intermediaries; this ensures experts accept integrity of the model; benchmarking against a suitable alternative (e.g., against logistic regression-derived scores); robustness in performance evaluation (e.g., effect of prevalence of different conditions); and comparative trials (Lisboa, 2002). Following these suggestions helps to enhance credibility in the development of CDSS tools.

10.2.1 BRIEF DESCRIPTION OF ARTIFICIAL NEURAL NETWORKS (ANNS)

ANNs use a process analogous to information processing by the human brain, acquiring knowledge through a learning process and storing it using inter-neuron connection strengths (synaptic weights). The ANN produces a set of outputs based on a presentation of input conditions. The ANN can be trained to estimate a specific outcome using the input variables in the database. ANNs are widely used and effective in a broad range of applications for analysis, control, and forecasting. The main advantage of neural networks over statistical techniques is that the model does not have to be defined prior to running the experiments. ANNs can recognize relevant data and patterns, whereas a statistical model requires prior knowledge of the relationships between the factors under investigation. Also, with statistics, it is difficult to integrate data with different formats such as continuous, binary, ordinal, and nominal data; but this can be achieved using ANNs.

Medical databases are generally large, containing more variables than are needed to predict the desired outcome. After the database has been cleaned (removal of outliers and ambiguous data), the first step is to pare down the database to a workable size by removing unimportant input variables. The difficulty lies in knowing which variables to remove. When only the most important variables remain, the minimum dataset has been identified. The minimum dataset defines the most important input variables for a prediction model; their relative importance can be listed by the weights of each input variable when the ANN has reached the point of maximum performance. When a database is mined, sometimes the ability to predict case-by-case results is not the goal. In some instances, a set of indicators may be desired in order to make some conclusions about the factors that lead to certain results. The variables of the minimum dataset (MDS) with the largest relative weights are these factors or indicators. See Figure 10.1 for an example of an artificial neural network with an input layer, one hidden layer, and an output layer, commonly referred to as a three layer network (Wikibooks_ANN).

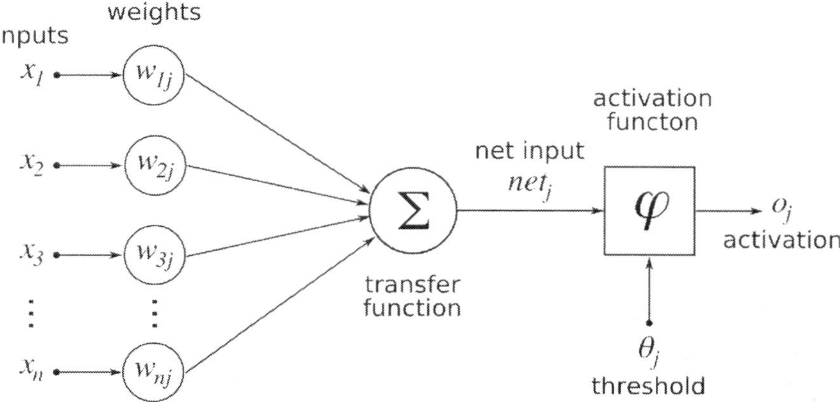

Figure 10.1: Artificial Neural Network with three layers. (Source: http://en.wikibooks.org/wiki/Artificial_Neural_Networks/)

Several methods have been proposed to identify the weights of the inputs and hidden nodes in order to establish the importance of each variable in predicting the outcome. One approach is to reduce the input variables one by one and observe when the performance deteriorates (Frize et al., 2001). The remaining variables then form a MDS to estimate a particular outcome. Another way is to add the weight-elimination algorithm when running ANN experiments, as this will send the small weights to zero and identify the remaining variables as those which best predict the outcome (McGowan et al., 1996).

There are several types of neural networks, but the most commonly used are: the feedforward-backpropagation algorithm, the multilayer perceptron neural network, fuzzy neural networks, and recurrent networks; more information on these approaches can be found in Micheli-Tzanakou, (2000); Haykin, (1996); Hudson and Cohen, (1999). There is also a popular book on genetic algorithms by Goldberg (2013) which is another pattern classification approach. This list is meant as a guide rather than a review of the literature on this vast topic.

Another method to eliminate variables is to use decision trees. Maintaining robustness in the performance of a DT classifier requires the creation of numerous decision trees (DTs), a process called ensemble classifiers. The ensemble classifiers are made by bootstrapping the original training sets and allocating these into bootstrapped training subsets. The DT classifier creates a tree from each training subset and tests each tree on the test set. A typical tree is created by using the best split among all variables. The growing and pruning algorithms are used to create a DT test to find the variables best predicting the selected outcome. The DT algorithm first classifies each class according to conditions, followed by the pruning steps until a global optimum is reached. The pruning portion creates a tree that is able to generalize well and deletes variables with low significance (Duda et al., 2001). Features appearing in the pruned DT become the selected subset of features that can be presented to an ANN for solving a classification problem (Dash and Liu, 1997; Frize et al., 2010). Once a tree is created, the variables are ranked according to the most significant variables and least significant variables are eliminated.

10.2.2 CASE-BASED REASONING (CBR) SYSTEM

A CBR system is an expert system that derives solutions to problems based on past cases. A basic premise in CBR is that many problems decision-makers encounter are not unique, but are mainly variations of a problem type. It is more efficient to solve a problem by analogy and starting with the solution to a previous similar problem than to generate the entire solution again from first principles (Morris, 1995). In solving a current problem, the CBR retrieves a similar past case and its solution. It then adapts the successful solution of the retrieved case to adjust for any differences between the current case and the retrieved case. Finally, the CBR stores the solution to the current case along with feedback about the outcome so that it can be used in solving future problems (Craw, 2005).

A CBR was created to match entire cases for inspection by physicians in an adult intensive care unit (ICU) and to generate warnings if any of the 10 closest matched patient cases have died (Frize and Walker, 2000; Taylor et al., 1993). It was suggested that a database could be used in conjunction with an ANN to impute missing values into medical databases (Ennett et al., 2001 and 2008; Ennett, 2003). The weights at the input nodes of an ANN are extracted when reaching an optimal performance and used in the CBR as match weights. Ten closest patient cases are found, and then missing values in the variables of these ten patients are imputed with the mean value of each variable for this set of 10 patients. There are other methods to impute missing values, such as replacing them with normal values. However the approach described above was found to perform as well or better than others described in the literature (Ennett et al., 2001 and 2008; Ennett, 2003).

REFERENCES

Ambalavanan, N, Carlo, WA (2001). "Comparison of the Prediction of Extremely Low Birth Weight Neonatal Mortality by Regression Analysis and by Neural Networks." *Early Hum. Dev.*, 65:123-137. DOI: 10.1016/S0378-3782(01)00228-6.

Craw, S (2005). "Case-Based Reasoning for Tablet Formulation Page." Available at: http://www.comp.rgu.ac.uk/staff/smc/papers/bpc01.pdf ; last accessed September 2013.

Dash, M, Liu, H (1997). Feature Selection for Classification. *Intelligent Data Analysis*. Elsevier: 131–156. DOI: 10.1016/S1088-467X(97)00008-5.

Duda, R, Hart, P, Stork, D (2001). *Pattern Classification.* 2nd ed. John Wiley & Sons, Inc., New York.

Ennett, CM, Frize, M, Walker, CR (2001). Influence of Missing Values on Artificial Neural Network Performance. *Proc. Medinfo'01 and British Library Direct ISSU* 84:449-453.

Ennett, CM (2003). "Imputation of Missing Values by Integrating Artificial Neural Networks and Case-Based Reasoning." PhD Thesis, Systems and Computer Engineering, Carleton University, Ottawa, ON, Canada.

Ennett, CM, Frize, M, Walker, RC (2008). "Imputing Missing Values by Integrating Neural Networks and Case-Based Reasoning." *Proc. IEEE/EMBS*, Vancouver: 4337-4341. DOI: 10.1109/IEMBS.2008.4650170.

Frize, M, Walker, R (2000). "Clinical Decision Support Systems for Intensive Care Units Using Case-Based Reasoning." *Med. Eng. Physics*, 22: 671-677. DOI: 10.1016/S1350-4533(00)00078-3.

Frize, M, Ennett, CM, Hebert, P (2001). "Improving the Efficiency and Effectiveness of Artificial Neural Networks as Decision-Support Systems." *Proc. Medinfo 01,* IOS Press, Amsterdam. DOI: 10.3233/978-1-60750-928-8-574.

Frize, M, Walker, RC, Catley, C (2007). "Healthcare Knowledge Management: Knowledge Management in the Perinatal Care Environment." Chapter 17 in: *Healthcare Knowledge Management: Issue, Advances, and Successes.* Eds. Bali RK, Dwivedi AN. Springer: 232-259. DOI: 10.1007/978-0-387-49009-0_17.

Frize, M, Weyand, SA, Bariciak, E (2010). "Suggested Criteria for Successful Deployment of a Clinical Decision Support System (CDSS)." *Proc. Med. Measure. Applic.* Ottawa: 69-72. DOI: 10.1109/MEMEA.2010.5480227.

Frize, M, Yu, N (2010). "Estimating Pre-Term Birth Using a Hybrid Pattern Classification System." *Proc. MEDICON 2010,* Thessaloniki, Greece: 893-896. DOI: 10.1007/978-3-642-13039-7_226.

Frize, M, Bariciak, E, Gilchrist, J (2013). "PPADS: Physician-PArent Decision-Support for Neonatal Intensive Care." *Proc. Medinfo* 2013: 23-27. DOI: 10.3233/978-1-61499-289-9-23.

Goldberg, DE (2013). *Genetic Algorithms.* Pearson Education.

Haykin, S (1996). "Neural Networks Expand SP's Horizons." *IEEE Signal Process. Mag.*, 13(2): 24-49. DOI: 10.1109/79.487040.

Hudson, DL, Cohen, ME (1999). *Neural Networks and Artificial Intelligence for Biomedical Engineering.* John Wiley and Sons. DOI: 10.1109/9780470545355.

Hyzg, RC (1995). "ICU scoring and clinical decision making." *Chest*, 107(2): 1482-1483. DOI: 10.1378/chest.107.6.1482.

Knaus, WA, Zimmerman, JE, Wagner, DP, Draper, EA, Lawrence, DE (1981). "APACHE – Acute Physiology and Chronic Health Evaluation: A Physiologically Based Classification System." *Crit. Care Med.* 9: 591-597. DOI: 10.1097/00003246-198108000-00008.

Lee, SK, McMillan, DD, Ohlsson, A, Pendray, M, Synnes, A, Whyte, R, et al., and the Canadian NICU Network (2000). Variations in Practice and Outcomes in the Canadian NICU Network: 1996-1997. *Pediatrics*, 106:1070-9. DOI: 10.1542/peds.106.5.1070.

Lisboa, PJG (2002). "A Review of Evidence of Health Benefit from Artificial Neural Networks in Medical Intervention." *Neural Networks*, 15: 11-39. DOI: 10.1016/S0893-6080(01)00111-3.

Meadow, W, Lantos, J (2003). "Ethics at the Limit of Viability: A Premie's Progress." *NeoReviews*, June, 4 (6): e157-e162. DOI: 10.1542/neo.4-6-e157.

McGowan, HCE, Stevenson, M, Frize, M (1996). "The Need for Standardized Reporting of Medical Applications of Artificial Neural Networks." *Proc. CMBEC*, Charlottetown, June.

Micheli-Tzanakou, E (2000). *Supervized and Unsupervised Pattern Recognition: Feature Extraction and Computational Intelligence.* CRC Press.

Morris, B (1995). "Case-Based Reasoning." West Virginia University. *AI/ES Update*, 5(1).

Richardson, DK, Gray, JE, McCormick, MC, Workmann, K, Goldmann, DA (1993). "Score for neonatal acute physiology: a physiologic severity index for neonatal intensive care." *Pediatrics*, 91: 617-623.

Richardson, DK, Corcoran, JD, Escobar, GJ, Lee, SK, and the Canadian Neonatal Network (2001). "Kaiser Permanente neonatal minimum data set area network, SNAP-II study group. SNAP-II and SNAPPE-II: Simplified newborn illness severity and mortality risk scores." *Pediatr*ics, 138: 92-100. DOI: 10.1067/mpd.2001.109608.

Richardson, DK, Tarnow Mordi, WO (1994). "Measuring illness severity in newborn intensive care." *J. Intensive Care Med.* 9:20 33. DOI: 10.1177/088506669400900104.

Sargent, DJ (2001). "Comparison of artificial neural networks with other statistical approaches – Results from medical data sets." *Cancer*, 91(8 Suppl S): 1636-1642. DOI: 10.1002/1097-0142(20010415)91:8+<1636::AID-CNCR1176>3.0.CO;2-D.

Taylor, KB, Nickerson, BG, Frize, M, Solven, FG, Dunfield, V (1993). "Use of Case-Based Reasoning to Assist Patient Management in an Intensive Care Unit." *CMBEC/CCGB Conference*, Ottawa: 248-249.

Wikibooks_ANN. Available at: http://en.wikibooks.org/wiki/Artificial_Neural_Networks/Print_Version; last accessed July 2013.

Weigend, AS, Rumelhart, DE, Huberman, BA (1990). "Back-propagation, weight-elimination and time series prediction." In: Touretzky, DS, Elman, JL, Sejnowski, TJ, Hinton, GE, Eds. *Proc. 1990 Connectionist Models Summer School. San Mateo: Morgan Kaufmann,*: 105-116.

CHAPTER 11

Knowledge Translation (KT), Integration, and Sharing (KIS) in a Clinical Environment

The chapter begins with the short definition of the last two steps in the KM process mentioned in Chapter 9. This is followed by a discussion on what physicians are looking for in Clinical Decision Support Systems (CDSSs) and their strengths and limitations. Since the example to illustrate these two steps is again taken from the perinatal experience, this clinical environment is described briefly. The examples include predicting premature births and delivery type for obstetrical patients and mortality of the infants after birth. The chapter ends with a discussion of the use of web services and web semantic services to interact with decision support systems and make these tools available where they are needed.

11.1 CLINICAL DECISION SUPPORT SYSTEMS (CDSS)

Clinical decision support systems (CDSS) have been developed and used for several medical applications. Kawamoto et al. (2005) did a systematic review of 70 studies from the literature to assess whether CDSSs can improve clinical practice; the authors concluded the following: decision support systems significantly improved clinical practice in 68% of the trials. Features that were found to be most significant were:

1. the decision support was provided automatically as part of the clinical workflow;

2. the decision support was provided at the time and location of decision-making;

3. the recommendations provided could be acted upon; and

4. they were computer-based rather than manual.

The authors also concluded that CDSSs show great promise for reducing medical errors and improving patient care. However, 34% of the studies did not show an improvement and the reason for their failure was not clear according to the authors (Kawamoto et al., 2005).

Many of the failures of early CDSSs occurred because the user had to filter the information and discard erroneous or useless information. This required much user time and the user had to interact with the system rather than just be a passive recipient of the output (Berner and La Lande, 2007; van Schiak et al., 2004). It is important that the CDSS be able to anticipate the need for information and deliver it in real-time without clinicians explicitly asking for it (van Schiak et al., 2004).

A study of the literature by Frize and Weyand (2010) identified criteria for a successful implementation of CDSSs in a clinical environment. The criteria for a successful deployment of a CDSS can be divided into three main areas: (i) the data entry and the effectiveness of algorithms used for the decision support; (ii) the human-computer interaction, including the data acquisition, and the manner in which information is requested from the system; and (iii) the output of the CDSS, including the format and type of information supplied (Frize and Weyand, 2010).

The data and information entry into the CDSS is one of the leading causes of failed CDSSs (Williams, 1992). Systems must require the least amount of physician time and be able to update themselves automatically (Bates et al., 2003). Some systems require users to enter patient data manually, which is time consuming and disrupts the delivery of patient care and the workflow of users. Manual data entry can be minimized by integrating the CDSS with the hospital information system and with electronic medical records. A successful CDSS will be able to extract data with limited user-interaction. Minimizing the time physicians spend entering data manually leads to greater usage and satisfaction with systems and helps to ensure time is not being taken away from the provision of patient care. Alternatively, if manual data entry is required, system implementation will be more successful if physicians are not expected to input this data themselves (van Schiak et al., 2004).

Another important issue is the maintenance of an up-to-date CDSS decision algorithm. Since patient management is constantly changing, the CDSS can easily become obsolete. CDSSs are often created with soft funding and when the funding runs out, keeping the system up-to-date becomes a major challenge. It is critical that systems be designed to have automated updating features. One way to do this is to have the CDSS retrain itself periodically and automatically (van Schiak et al., 2004; Gago and Santos, 2008; Frize and Weyand, 2010; Ong et al., 2013).

Richardson and Ash (2011) surveyed how physicians viewed CDSSs; physicians saw CDSSs as software tools that can provide alerts, prompts, and references, but not as tools supporting patient management, clinical operations, or workflow, all features these physicians said they would like to have. Physicians also wanted CDSSs that enhance the physician-patient relationship, help redirect work among staff, and save them time. The authors claimed that participants in their survey were generally dissatisfied with current CDSS capabilities and also with the electronic health record usability. Physicians identified the following aspects of decision-making that they would like to have: medication administration and treatment, and cognitive decision-making that enhance relationships and interactions with patients and staff. Physicians expressed a need for decision support that integrates functions in a way that aligns time and resources and assist providers in a broad range of decisions (Richardson & Ash, 2011).

A panel at a Human Factors and Ergonomics Society Annual Meeting discussed what contributions human factors science could bring to bear on (1) the design, (2) integration, and (3) training necessary for effective CDSS implementation. Their conclusions were similar to those of Bates et al. (2003): members of the panel suggested that CDSSs are effective in enhancing clinical performance for drug dosing, preventative care, and diagnostic accuracy, but that they must improve significantly because of poor usability, slow processing and display speed, lack of natural language processing, and a lack of proper task analysis/integration which limited their impact on important patient safety indicators. In order for CDSSs to achieve widespread acceptance and greater efficacy, empirically derived guidelines must be applied in their design and implementation. Further, the resultant systems must be proven effective in a variety of medical contexts, from high-stress emergency room settings to day-to-day procedures where accurate and timely data can save lives (Karsh et al., 2010).

Ash et al. (2012) studied 34 community hospitals in the U.S. which had 5 years of experience using computerized provider order entry (CPOE), to assess their standard practices. They found that using CPOE with clinical decision support systems (CDSSs) can help hospitals to improve care. The hospitals uniformly used medication alerts and order sets, had sophisticated governance procedures for CDSSs, and employed staff to customize these systems. However, they found that the level of customization needed for most CDSSs before they could be fully implemented was greater than what they had expected. Customization required skilled individuals, an emerging specialization. The authors of the study concluded that the results bode well for a robust diffusion of CDSS for similar hospitals and suggested that national policies to promote the use of CDSSs may be successful (Ash et al., 2012).

In an editorial, Loback (2013) asked the question: "The road to effective clinical decision support: are we there yet?" He responded:

> "No, we are only beginning to understand what factors make decision support effective and we may need to chart a new course...The field should embrace a continuous quality improvement approach through which real world field based observations on the use of decision support, combined with qualitative contextual evaluation, can inform developers and users about what does and does not work. Accordingly, decision support systems should be increasingly created to allow this type of real time monitoring and evaluation. So for now, the journey toward what constitutes effective clinical decision support is far from over." (**Lobach, 2013**).

Seidling et al. (2011) report on a novel approach to predict the success of clinical decision support and identify factors that influence the acceptance of alerts. The type of alert they looked at was drug interaction alerts. The authors argue that clinical decision support has the potential to improve care, but they often have low user-acceptance rates. Some of the factors affecting whether or not decision support is accepted appear to include workflow issues, the intrusiveness and importance of the alert, and for who it was displayed. In their study, Seidling et al. (2011) attempted to quantify the impact of some of the major human factors issues which may affect alert acceptance by specification of three variables: the display of the alert, the textual information, and the prioritization of the alerts; they found that the alert display most strongly correlated with alert acceptance, reinforcing that the presentation of the alerts at the user interface is an important determinant of alert acceptance. To optimize current and future systems, the authors mention the importance of considering human factors and esthetics as well as the psychological aspects of human-computer interaction in the development process (Seidling et al., 2011).

11.1.1 DESIGN OF A CDSS FOR THE OBSTETRICAL ENVIRONMENT

Perinatal care includes the period during the pregnancy, and the care given to newborns after delivery. The period encompasses the 28th week of gestation through the 7th day after delivery. Data collection from the mother and fetus occurs during different stages of the pregnancy and it continues with the infant after the birth, especially if there are health issues. The type of data collected depends on the clinical needs and research plans for data utilisation and prediction models to be derived from the data. Integrating data that comes from so many sources, in an intelligent manner, provides knowledge that should help caregivers in making a diagnosis and in selecting a course of therapy. Caregivers also need to estimate outcomes or whether severe complications are likely to arise.

The results of the data mining and of artificial intelligence tools can be used in CDSSs or in electronic medical records (EMRs) and should be integrated into the clinical workflow to create value to support healthcare decisions and provision. This means that the manner in which knowledge is accessed and viewed must be user-friendly and require minimal computer interaction by the caregivers. This step is illustrated by an example of decision-support in the neonatal intensive care environment (NICU). In the NICU, for example, accurate outcome predictions have the potential to improve clinical outcomes, facilitate patient management, aid in parental counseling, and in planning resource allocation (Frize et al., 2013).

From conception to birth, a vast quantity of patient information relating to an infant and the mother is accumulated and stored in medical databases. Infant-related data are stored in obstetrical, perinatal, and the neonatal intensive care for very sick infants; these databases are frequently distributed temporally and geographically. When developing systems for outcome estimation, it is essential to find ways to reduce the number of variables used to estimate outcomes; a minimum dataset (MDS) should be defined for each of the outcomes to be estimated. Moreover, there is a need to obtain faster test results and to predict the onset of significant or life-threatening changes. This need, coupled with the limitation of current illness severity indices, has been a motivating force behind research and development efforts to create clinical decision-support tools (CDSSs) for a perinatal environment. Figure 11.1 shows the outcomes that can be estimated at various stages of perinatal care, both during the pre and post-natal period. These outcomes are thought to be highly relevant for obstetricians, neonatologists, and pediatricians and are discussed in more detail below.

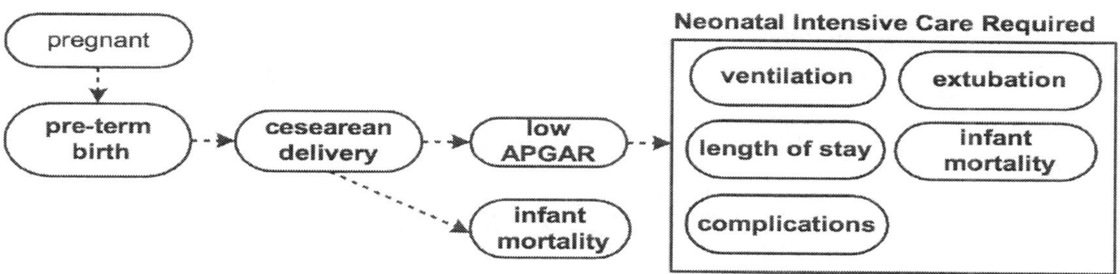

Figure 11.1: Types of data collected in a perinatal environment for analysis. (Source: Frize et al. 2007; also adapted from Catley 2007)

Estimating Pre-Term Births (PTB)

Pre-term birth is defined as birth before 37 completed weeks of gestation. PTB is the leading cause of mortality occurring before 28 days of age, accounting for 85% of all neonatal deaths not due to lethal congenital malformations (Dawood, 1980). Due to its direct correlation with infant mortality and increased risk of long-term health problems, reducing the burden of PTB has become the number one neonatal health priority. Many studies have attempted to predict women at risk of PTB; so far, no scoring system has proven itself superior to clinical judgment. One of the major obstacles is that many women who deliver prematurely have no obvious risk factors; over half of all PTBs occur in low-risk pregnancies (Iams et al., 2001; Allan et al., 2002). Due to a lack of complete obstetric data, it is often difficult to perform risk estimation based on sophisticated medical markers and extensive medical histories, as published in large clinical trials (Goldenberg et al., 1998).

A study used Artificial Neural Networks (ANNs) to predict PTB by classifying each new patient case and identifying mothers at high-risk of delivering premature infants (Catley et al., 2005; Frize and Yu, 2010). The hybrid DT-ANN (decision tree to reduce variables and ANN to estimate PTB) produced a test sensitivity (true positive cases of PTB) of 66.0% for parous women (not a first pregnancy). The best test specificity (true negative cases) was 80%. Each classifier had a similar area under the Receiver Operating Characteristic (ROC) curve values of 0.80, indicating excellent discrimination between women at risk of PTB and those who were not at risk of delivering prematurely. For nulliparous women (first pregnancy), the performance of the prediction was slightly lower as expected, since an important predictor is the occurrence of a previous pre-term birth; for these cases, the best result was a test sensitivity of 65.0%, and a test specificity of 71.3%. These results are somewhat higher than those reported in the literature to date for a general population and nulliparous women. The ROC was 0.71 for the latter. All classifiers in the parous and nulliparous cases had verification results that closely matched the test set results. The positive predictive values (PPV) for parous cases were greater than 40%, with a high value of 44.2%; the highest Negative Predictive Value (NPV) was 92.9%. The best nulliparous PPV was 32.3%; the NPV of 90.6%, a small difference of 2.2% from the parous NPV (Frize and Yu, 2010).

Delivery Type

The Caesarean birth rate has been increasing slowly every year in Ontario (Canada), from 19.9% in 1999, to 20.9% in 2000, and 22.8% in 2001. The increase occurred in both teaching and large community hospitals. The debate continues on whether the present caesarean birth rate is too high. A study of 593 hospitals in the U.S., with data collected in 2009, found that Caesarian rates varied between 7.1 and 69.9%; and for women at low risk, the variation was between 2.4 and 36.5%. The authors called this situation a costly overuse of C-section delivery (Kozhiminnil et al., 2013).

The outcome "delivery type" predicts whether the delivery of the baby will be by Caesarian section or a vaginal delivery. Successfully predicting the delivery type can help to prepare for the care of mothers during delivery. In a study of the development of a prediction model for delivery type (C-section or vaginal delivery), two different artificial intelligence approaches were used: Artificial neural networks and Fuzzy-logic neural networks. Both methods resulted in a high performance measured by the sensitivity, specificity, and ROC. The sensitivity was 84.0% for the three-layer network (i.e., with one hidden layer). The specificity was 86.2% in the two and three-layer networks. With the Fuzzy neural network, the sensitivity was 71.3% and the specificity was 94.3%. These results show a very good performance of both approaches in estimating delivery type (Frize et al., 2004).

Apgar score

The Apgar test is done one minute and five minutes after delivery to evaluate a newborn's condition immediately after birth. The Apgar tests five attributes: appearance (color); pulse (heartbeat); grimace (reflex); activity (muscle tone); and respiration (breathing). A score is determined by awarding zero, one or two points in each category. Scores of seven and higher indicate that the baby is in good clinical condition. The Apgar score was developed in 1952 by the late pediatrician, Dr. Virginia Apgar. Again, a prediction model was developed for this rarely studied outcome with encouraging results. With the ANN, the specificity was very high (99.7%), but the sensitivity was lower (31.4%). This particular outcome does not appear to have been studied to date by artificial intelligence tools. This first attempt is encouraging and the ANN will be optimized by a variety of approaches to obtain a higher specificity in the future. With the Fuzzy neural network, the sensitivity was 19.5%, and the specificity was 99.3% (Frize et al., 2004).

11.1.2 ESTIMATING NEONATAL INTENSIVE CARE OUTCOMES

Development of neonatal technology and practices have resulted in sophisticated care, capable of sustaining life for infants as small as 500 grams and as premature as 23 weeks gestation. There has been a substantial reduction of mortality in premature infants, and the rate of handicap or significant morbidity appears to have remained steady or to have declined in survivors of NICUs of nearly all gestational ages and weights. Despite these positive results, current neonatal care decisions are frequently based on uncertain conditions, and the use of aggressive treatments is increasingly being questioned. A serious concern for health care providers and for parents is: To whom should this intensive care be administered and in what circumstances should it be withdrawn? Are there factors with respect to these babies' health status to guide physicians and parents on whether to administer treatment or not, or to end it if it has been started? In the NICU, ethical dilemmas and conflicts in decision-making are unavoidable. The challenge is to provide guidance that is practical, specific, but not prescriptive. If estimates can be provided with acceptable accuracy, the potential exists to significantly enhance the evidence on which critical decision-making is based in NICUs (Frize et al., 2005; Weyand et al., 2011; Frize et al., 2013).

Mortality

Neonatal mortality is defined by the World Health Organization as death occurring before 28 days of age. Earlier research used data collected by the Canadian Neonatal database (CNN), where only deaths to discharge were recorded, because there was no follow up after discharge from the hospital. Moribund babies were excluded from the database since they only receive comfort measures rather than aggressive therapy.

In a study that used ANNs, the Positive Predictive Value (PPV) was 62.7% for infants weighing less than 1000 g. The ANN system consistently demonstrated high specificity and Negative Predictive Value (NPV) suggesting that it was relatively accurate in predicting survival. This is an important finding for clinical utility. Caution must be exercised in predicting death, as this might lead to inappropriate counseling of parents and subsequent inappropriate ethical and therapeutic decision-making. Given that caregivers predict death quite well early in the course of a NICU patients' stay, but predict survival less well, the ANN might be an excellent complement to clinical judgment, since it appears to predict survival more accurately than caregivers. Nonetheless, improving sensitivity and positive predictive value for the outcome of mortality is an important aspect to improve the potential utility of these systems for use in evidence-based ethical decision-making (Walker and Frize, 2004).

A more recent study of mortality of infants in the NICU used real-time data collected from 20 infants in the NICU at the Children's Hospital of Eastern Ontario (CHEO). The best performance was obtained when using data of the first 48 h after admission; the Decision Trees (see 5) estimated mortality with an average sensitivity of 18% and an average specificity of 99%; the positive predictive value (PPV) was 57%, and the negative predictive value (NPV) was 95%. The ANN mortality estimation, using the ten 5by2 cross validation data sets, had an average sensitivity of 63% with a Confidence Interval (CI) of 0.06, an average specificity of 99% (CI 0.00), an average positive predictive value (PPV) of 73% (CI 0.06), and an average negative predictive value (NPV) of 98% (CI 0.00). The average F1 score was 0.67 (CI 0.04), and the average Matthews Correlation Coefficient (MCC) was 0.66 (CI 0.04). According to the medical experts involved in these studies, a clinically useful mortality prediction should have a sensitivity of 60% or greater, and a specificity of 90% or larger. The mortality model in this study surpassed the criteria for clinically useful results (Frize et al., 2013).

Other NICU Outcomes

Other outcomes studied in the NICU were: estimating the duration of artificial ventilation, i.e., whether the ventilator would be used for 12, 24, 48 h 7 days, 14 days, or longer, and length of stay (Townsend and Frize, 2008). Another study estimated whether extubation would succeed or fail (Mueller et al., 2004). Potential clinical complications of neonates in the intensive care unit are currently being studied with encouraging results; examples are: major neuro-imaging abnormality (NIA); chronic lung disease (bronchopulmonary dysplasia-BPD), necrotising enterocolitis (NEC), and retinopathy of prematurity (ROP).

11.2   KNOWLEDGE TRANSLATION TO USERS:

In this step, it is important to present the knowledge in a manner which is intuitive and user-friendly. Users can be caregivers or patients themselves; the latter would include parents or guardians for children or for persons who are not mentally capable of making decisions. An example developed for the NICU follows.

The design consists of a module for the clinicians, connected to a module for parents of infants in the NICU. The modules can be customized based on the need or preference of users. One module contains an overview of important information on all the babies in the unit that the clinician can quickly review at a glance. This page includes a list of recent alerts identifying patients whose outcome predictions are in the high or medium risk range, and those who have had significant changes to their alert ranges in either direction. The CDSS has the capability of generating intelligent alerts and warnings on the monitor screen; these can also be sent to a handheld or mobile device. The CDSS predicts mortality and models are being developed to predict NEC, IVH, BPD, long term ventilation and length of stay in the NICU using real-time data. Clinicians can access the most recent prediction outcomes, as well as previous trends, and assess how patient risk-levels have changed in response to the treatments administered. Clinicians can also access historical patient information contained within the system which can be searchable by patient name, hospital number, or by dates. Authorized clinicians have the option to edit the information already in the system and to add new explanations, definitions, and notes. Since the system is modular, more modules can be added as desired.

The Parent Module contains a subset of the options provided in the Clinician Module and has restricted permissions. The Parent Module associated with each patient file is activated by a physician through the Clinician Module. This activation involves pre-selecting the decision-support modules to be made available to parents; the default consists of the medical explanations and definitions and a unique username and password for the parents to log-in. The system monitors the progress a parent makes in each session towards a collaborative decision with physicians on a change of care. This information can be saved and resumed at a later time. A summary page covering the completed decision-support modules and any additional information can be printed in advance of a meeting with the physician or for personal review, following one of these sessions with the system (Weyand et al., 2011; Weyand, 2011).

The decision modules are designed to help parents make difficult treatment decisions such as withdrawal of treatment for the sickest babies in the NICU, a do not resuscitate (DNR) order, moving to palliative care, or limiting escalation of care. If enabled by the physician, the risk levels predicted for the various medical conditions provide parents with specific information about the condition of their infant. The parents can select any of the conditions pertaining to their infant and access definitions on each of the conditions. The module asks parents questions to help them consider the risks and benefits of the treatment and of the alternatives, and the relative importance of the options, benefits, harms, while considering their personal values. The questions and presentation style to parents follow the IPDAS (International Patient Decision Aid Standard). The material is presented at an eighth-grade level to facilitate a thorough understanding by parents with various educational backgrounds. The system can provide a more detailed explanation by selecting the option "Read More" (Weyand, 2011; Weyand et al., 2011; Frize et al., 2013). The usability pilot study of these decision support tools is discussed in the next chapter (12).

11.3   KNOWLEDGE INTEGRATION AND SHARING

Integration is a pressing concern for the health care domain, since there are so many sources of data with little connection between them. The data collected and the knowledge extracted are complex. Decisions must frequently be made in a fast-paced environment such as an emergency department or an intensive care unit. The Semantic Web and Web services are two emerging and evolving technologies with the potential to change the face of health care delivery, ultimately providing integrated healthcare data, knowledge, and computer-based applications such as decision-support tools and electronic medical records (Lea and Vinoski, 2003; Berners-Lee et al., 2001; Kwon, 2003; Catley et al., 2005; Catley, 2007).

Like conventional web services, Semantic Web Services (SWS) consist of the server end of a client-server system for machine-to-machine interaction via the World Wide Web. Adding semantics to the Web is a necessary component to provide a Web service that integrates various parts of the information needed for the effective provision of patient care. The goal of the Semantic Web is to associate meaning to web resources (Berners-Lee et al., 2001), making the Semantic Web a meaningful indexed repository of documents and services (Lee et al., 2004). As decision support system environments are rapidly changing from centralized and closed to distributed and open, scalability and interoperability features are becoming more crucial to CDSS development (Kwon, 2003). Combining the Semantic Web and Web Services can provide physicians and other caregivers instant access to knowledge instead of just raw data. To complete the KM cycle, the data, knowledge and services must be accessible to all practitioners involved in patient care in the medical unit for which they are designed.

An example below describes how Semantic Web Services have been used for a healthcare application, as a technological enabler, to create an infrastructure for perinatal outcome estimations. The infrastructure integrates both clinical data and clinical decision-support tools in a distributed healthcare environment. The objective was to autonomously compose Semantic Web Services based on pre-defined physician service composition templates. By formulating the physician's decision-making process, it is possible to alert physicians to potential adverse perinatal outcomes before they occur (Catley et al., 2005; Catley, 2007).

In the distributed field of perinatal care, developing models for outcome estimation, using minimum data sets (defined in Chapter 10), involves integrating patient cases from obstetrical, perinatal and neonatal care databases (see Figure 11.1). An ontology that covers relevant perinatal care and clinical decision-support terminology was created, essentially a predefined taxonomy for perinatal clinical decision-support. A perinatal CDSS ontology is a prerequisite to integrate data that are collected from distributed databases, providing the standardization needed to handle differences in terminology that occur when multiple databases represent the same information, but in different ways (Catley, 2007; Frize et al., 2007).

The tools described earlier, such as ANNs for outcome estimation, and a CBR system for matching an individual patient's condition to similar past cases, are key components of the particular infrastructure designed by Catley (2007). The information technology tools used to achieve integration are based on the World Wide Web Consortium (W3C) standards; they involve the use of XML to support standardized data, and Web services to offer decentralized clinical decision-support (Frize et al., 2007; Catley, 2007).

The infrastructure proposed by Catley (2007) has three key concepts, shown by the three layers depicted in Figure 11.2 and is discussed in Frize et al. (2007).

Layer 1 (*Knowledge Description*) encompasses the knowledge sources required by the higher layers, that is, the clinical perinatal databases and the practitioner's tacit domain knowledge. Each knowledge source must be formalized before being invoked by the Semantic Web Services: A medical ontology describes the database schemas, and composition templates model the manner in which core services are composed to perform complex service compositions.

Layer 2 (*Integration*) involves developing minimum data sets (MDSs) for perinatal outcome prediction. The MDSs are then integrated with the 'Generic Clinical Decision-Support Systems (CDSS) Workflow model for perinatal outcome estimation', which defines how the CDSs are composed as Semantic Web Services.

Layer 3 (*Knowledge Sharing*) provides the Semantic Web Services, which integrate the inner layers and the means by which the practitioner accesses the perinatal outcome estimation tools. When practitioners wish to interact with the system, they enter information on a new patient case through a Web services user-interface; data is obtained either directly from the patient or from a clinical repository. The practitioner can then perform outcome estimation and receive a result in real-time. The result is evaluated, possibly discussed with the patient, and ideally stored in the patient's Electronic Medical Record (EMR).

The three layers are interrelated: layer 2's objective cannot be met until layer 1's knowledge sources are described and obtained; the trained outcome estimation tools developed in layer 2 cannot be ubiquitously invoked, shared and evaluated until the Semantic Web Services infrastructure described in layer 3 is designed and implemented (**Frize et al., 2007; Catley, 2007**). See **Figure 11.2**.

Although there has not been as much development work and research done on the application of artificial intelligence to the perinatal care environment as for adult care, computers are becoming more prevalent in health care and many applications to support complex decision-making are being developed and tested. Very good quality databases are being compiled with physiologic data acquired from monitors, laboratory tests and medical images, electronic medical records, and hospital patient information systems; these are increasingly available in real-time. Thus, data mining tools and decision-aid tools with intelligent alerts and warnings will become increasingly available to several medical specializations in the future. However, while the provision of better 'evidence' to support clinical, 'ethical' and resource decisions appears to be a valuable contribution to care and decision-making, clinical trials of such systems must be done to ensure that their design is optimal, that they fit in the workflow of users, and lead to beneficial clinical applications. In the future, teams of developers must assess not only the performance of the systems, but also their impact on clinical practice, a topic which is discussed in the next and last chapter.

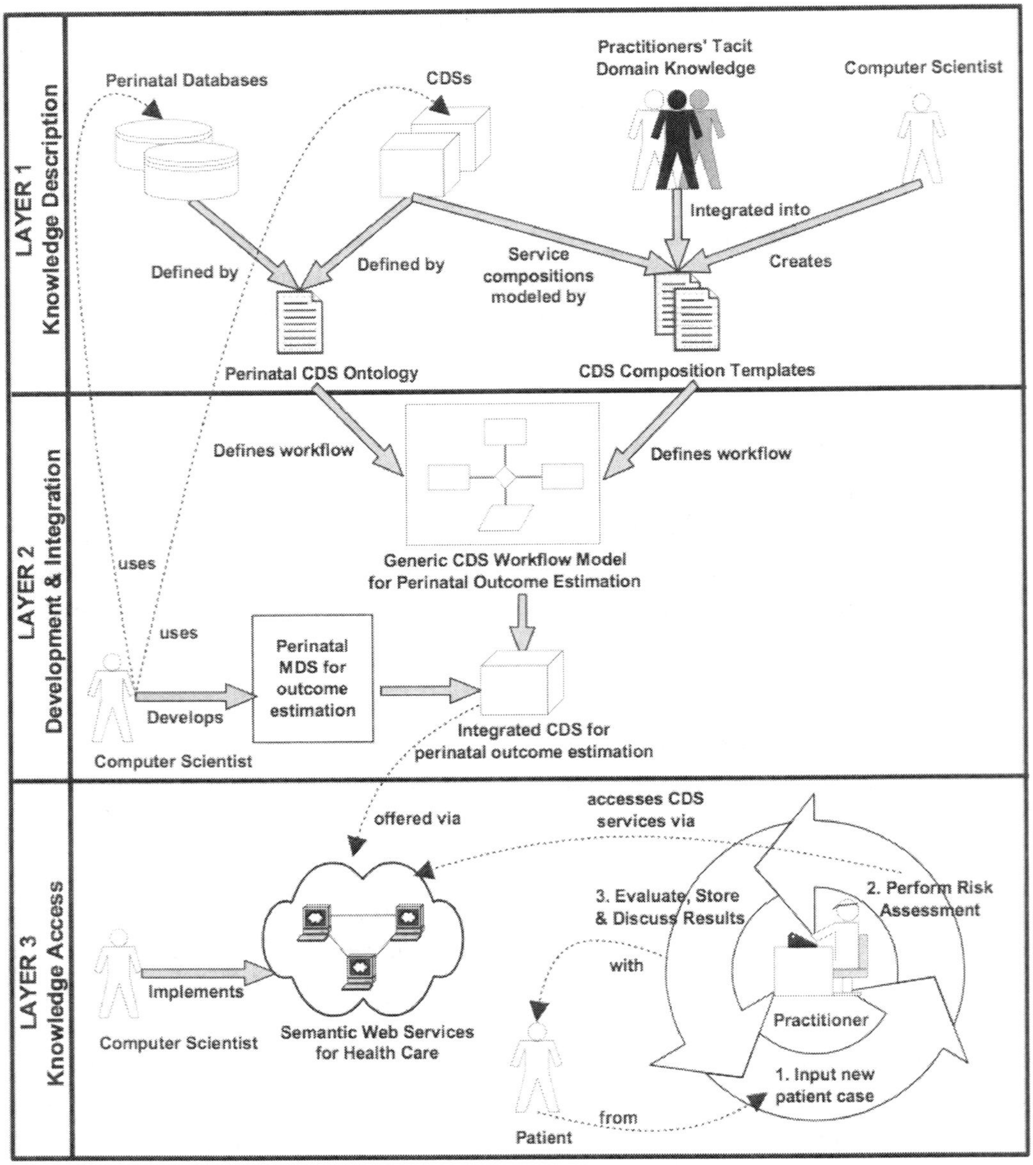

Figure 11.2: Knowledge description, integration, and sharing. (Source: Frize et al., 2007)

REFERENCES:

Allan, N, Aylward, D, Berry, E, et al. (2002). "Preterm Birth: Making a Difference." *Best Start Resource Centre,* Toronto, Canada.

Ash, JS, McCormack, JL, Sittig, DF, Wright, A, McMullen, C, Bates, DW (2012). "Standard Practices for Computerized Clinical Decision Support in Community Hospitals: A National Survey." *JAMIA,* 19(6): 980-987. DOI: 10.1136/amiajnl-2011-000705.

Bates, DW, Kuperman, GJ, Wang, S, Gandhi, T, Kittler, A, Volk, L, et al. (2003). "Ten Commandments for Effective Clinical Decision Support Making the Practice of Evidence Based Medicine a Reality." *JAMIA,* 10(6): 523-530. DOI: 10.1197/jamia.M1370.

Berner, ES, La Lande, TJ (2007). "Overview of Clinical Decision Support Systems" in *Clinical Decision Support Systems: Theory and Practice.* 2nd ed. Brimingham: ch. 1: 1-18. DOI: 10.1007/978-0-387-38319-4.

Berners-Lee, T, Hendler, J, Lasilla, O (2001). "The Semantic Web." *Scientific American*: 34-43. Available at: http://www.sciam.com/article.cfm?articleID=00048144-10D2-1C70-84A9809E-C588EF21. Accessed. DOI: 10.1038/scientificamerican0501-34.

Catley, C, Frize, M, Walker, CR, Petriu, DC (2006). "Predicting High-Risk Pre-Term Birth Using Artificial Neural Networks." *IEEE Trans. Inf. Tech. Biomed.* 10(3): 540-549. DOI: 10.1109/TITB.2006.872069.

Catley, C, Petriu, DC, Frize, M (2005). "A UML Framework for Web Services-Based Clinical Decision Support." *Proc. 14th Intern. Conf. on Intelligent and Adaptive Systems and Software Engineering.* Toronto, Canada.

Catley, C, Frize, M, Petriu, DC (2005). "Predicting Preterm Birth Using Artificial Neural Networks." *Proc IEEE Computer–Based Medical Systems.* Dublin, Ireland: 103-108. DOI: 10.1109/CBMS.2005.84.

Catley, C, Frize, M, Petriu, D, Walker, RC, Yang, L (2004). "Towards a Web Services Infrastructure for Perinatal, Obstetrical, and Neonatal Clinical Decision Support." *Proc. IEEE EMBS Conf.,* San Francisco: 3334-3337. DOI: 10.1109/IEMBS.2004.1403937.

Catley, CA (2007). "An Integrated Hybrid Data Mining System for Preterm Birth Risk Assessment Based on a 'Semantic Web Services for Healthcare' Framework." PhD Thesis, Systems and Computer Engineering, Carleton University, Ottawa, Canada K1S 5B6.

Dawood, MY (1980). "The Obstetric View of Premature Labor. In: Smith GF, Vidyasagar D, editors. Historical Review and Recent Advances in Neonatal and Perinatal Medicine." *Mead Johnson Nutritional Division.* Available at: http://www.neonatology.org/classics/mj1980/ch10.html; last accessed September 2013.

Frize, M, Yu, N (2010). "Estimating Pre-Term Birth Using a Hybrid Pattern Classification System." *Proc. MEDICON2010*, Thessaloniki, Greece, May: 893-896. DOI: 10.1007/978-3-642-13039-7_226.

Frize, M, Ibrahim, D, Seker, H, Walker, RC, Odetayo, MO, Petrovic, D, Naguib, RNG (2004). "Predicting Clinical Outcomes for Newborns Using Two Artificial Intelligence Approaches." *Proc. IEEE EMBS Conf.*, San Francisco: 3202-3205. DOI: 10.1109/IEMBS.2004.1403902.

Frize, M, Yang, L, Walker, RC, O'Connor, AM (2005). "Conceptual Framework of Knowledge Management for Ethical Decision-Making Support in Neonatal Intensive Care," *IEEE Transactions on Information Technology in Biomedicine*, vol. 9, pp. 205-215. DOI: 10.1109/TITB.2005.847187.

Frize, M, Walker, RC, Catley, C (2007). "Healthcare Knowledge Management: Knowledge Management in the Perinatal Care Environment." Chapter 17 in: *Healthcare Knowledge Management: Issue, Advances, and Successes*. Eds. Bali RK, Dwivedi AN. Springer: 232-259.

Frize, M, Weyand, SA, Bariciak, E (2010). "Suggested Criteria for Successful Deployment of a Clinical Decision Support System (CDSS)." *Proc. Med. Meas. Applications*, Ottawa: 69-72. DOI: 10.1109/MEMEA.2010.5480227.

Frize, M, Weyand, S, Gilchrist, J, Bariciak, E, Dunn, S, Tozer, S (2011). Combined Physician-Parent Decision Support Tool for the Neonatal Intensive Care unit. *Proc. MeMeA*, Bari, Italy. May: 59-64. DOI: 10.1109/MeMeA.2011.5966652.

Frize, M.Bariciak, E, Gilchrist, J (2013). "PPADS: Physician-PArent Decision-Support for Neonatal Intensive Care." *Medinfo 2013 Conference*, Copenhagen: 23-27. DOI: 10.3233/978-1-61499-289-9-23.

Gago, P, Santos, MF (2008). "Towards an Intelligent Decision Support System for Intensive Care Units." *18th European Conference on Artificial Intelligence: Workshop on Supervised and Unsupervised Ensemble Methods and their Applications.*

Goldenberg, RL, Iams, JD, Mercer, BM. et al. (1998). "The Preterm Prediction Study: The Value of New vs. Standard Risk Factors in Predicting Early and All Spontaneous Births." *Amer. J. Public Health*, 88(2): 233-238. DOI: 10.2105/AJPH.88.2.233.

Iams, JD, Goldenberg, RL, Merber, BM. et al. (2001). "The Preterm Prediction Study: Can Low-Risk Women Destined for Spontaneous Preterm Birth be Identified?" *Gen. Obst. Gynecol.*, 184(4): 652-655. DOI: 10.1067/mob.2001.111248.

Kozhiminnil, K (April 2013). "Higher Rates of Obstetric Intervention for the Privately Insured." Available at: http://www.advances.umn.edu/2013/04/study-shows-higher-rates-of-obstetric-intervention-among-privately-insured-women/; last accessed July 2013.

Kwon, OB (2003). "Meta Web Service: Building Web-Based Open Decision Support System Based on Web Services." *Expert Sys. Appl.*, 24: 375-389. DOI: 10.1016/S0957-4174(02)00187-2.

Karsh, BT, Guerlain, S, Metcalf, D, Marquard, JL, Gorman, P (2010). "Just What the Doctor Ordered?: The Role of Cognitive Decision Support Systems in Clinical Decision-Making & Patient Safety." *Proc. Human Factors and Ergonomics Society Annual Meeting*, 54(12): 826-829. DOI: 10.1177/154193121005401204.

Kawamoto, K, Houlihan, CA, Lobach, DF, (2005). "Improving Clinical Practice Using Clinical Decision Support Systems: A Systematic Review of Trials to Identify Features Critical to Success." *BMJ*, 330(7494): 765-777. DOI: 10.1136/bmj.38398.500764.8F.

Lea, D, Vinoski, S, (2003). "Middleware for Web services." *IEEE Internet Comput.* 7: 28-29. DOI: 10.1109/MIC.2003.1167336.

Lee, Y, Patel, C, Chun, SA, Geller, J, (2004). "Compositional Knowledge Management for Medical Services on Semantic Web." *Proc. 13th Intern. World Wide Web Conf.*, New York: 498-499. DOI: 10.1145/1013367.1013544.

Lobach, D, (2013). "The Road to Effective Clinical Decision Support: Are We There Yet?" *BMJ*, 346: Mar13_1 f1616. DOI: 10.1136/bmj.f1616.

Mueller, M, Wagner, CL, Annibale, DT, Hulsey, TC, Knapp, RG, Almeida, JS, (2004). "Predicting Extubation Outcome in Preterm Newborns: A Comparison of Neural Networks with Clinical Expertise and Statistical Modeling." *Pediatr. Res.* 56(1): 11-18. DOI: 10.1203/01.PDR.0000129658.55746.3C.

Ong, DE, Frize, M, Gilchrist, J, Bariciak, E, Ennett, CM, (2013). "Usefulness Analysis of a Clinical Data Repository Design." *Proc IEEE Medical Measurements and Applications (MeMeA)*. DOI: 10.1109/MeMeA.2013.6549712.

Richardson, JE, Ash, JS, (2011). "A Clinical Decision Support Needs Assessment of Community-Based Physicians." *JAMIA*, 18: Suppl_1 i28-i35. DOI: 10.1136/amiajnl-2011-000119.

Seidling, HM, Phansalkar, S, Seger, DL, Paterno, MD, Shaykevich, S, Haefeli, WE, Bates, DW, (2011). "Factors Influencing Alert Acceptance: A Novel Approach for Predicting the Success of Clinical Decision Support." *JAMIA*, 18(4): 479-484. DOI: 10.1136/amiajnl-2010-000039.

Townsend, D, Frize, M, (2008). "Complementary Artificial Neural Network Approaches for Prediction of Events in the Neonatal Intensive Care Unit." *Proc IEEE/EMBS Conf.* 4605-4608. DOI: 10.1109/IEMBS.2008.4650239.

Van Schiak, P, Flynn, D, Van Wersch, A, Douglass, A, Cann, P, (2004). "The Acceptance of a Computerised Decision-Support System in Primary Care: A Preliminary Investigation." *Behav. Infor. Tech.*, 23(5): 321-326. DOI: 10.1080/01449290041000669941.

Walker, RC, Frize, M, (2004). "Are Artificial Neural Networks 'Ready to Use' for Decision-Making in the NICU?" *Paediat. Res.*, 56:6-8. DOI: 10.1203/01.PDR.0000129654.02381.B9.

Weyand, SA, Frize, M, Bariciak, E, Dunn, S, (2011). "Development and Usability Testing of a Combined Physician-Parent Decision-Support Tool for the Neonatal Intensive Care Unit." *Proc. IEEE/ EMBC*, Boston: 6430-6433. DOI: 10.1109/IEMBS.2011.6091587.

Weyand, S, (2011). "Development and Usability Testing of a Neonatal Intensive Care Unit Physician-Parent Decision Support Tool (PPADS)." MASc Thesis, Systems and Computer Engineering, Carleton University, Ottawa, ON, Canada.

Williams, LS, (1992). "Microchips versus Stethoscopes: Calgary Hospital, MDs Face off over Controversial Computer System." *CMAJ*, vol. 147, pp. 1534-40, 1543, 1992.

OTHER SUGGESTED READING

Bali, RK, Dwivedi, AN, Eds. (2007). *Healthcare Knowledge Management: Issues, Advances, and Successes.* Health Informatics Series. Springer. DOI: 10.1007/978-0-387-49009-0.

CHAPTER   12

Clinical Trials and Usability Studies in a Medical Environment

This chapter presents some elements of two approaches for the validation of a new technology or of a new software tool. Clinical trials are essential prior to marketing a new technology or software tool for use on humans. This applies to the testing of a new software tool, a medical device, a modification of an existing technology, or of a new medical procedure. Clinical trials are costly and usually involve a fairly substantial number of participants. When a prototype is at an earlier stage of development, perhaps at the proof-of-concept stage, there should be a short pilot study with a smaller number of participants. A usability study, especially if this is for a software tool or a new device, is usually carried-out first in a laboratory where it is designed, and then in a location where it is meant to be used in the future. This approach can be repeated each time that the tool or device is modified, until a final prototype is ready for the fuller clinical trial stage.

Every new technology development, or a modification of a technology, intended for use with humans, must be tested appropriately prior to its dissemination and clinical use. However, there are various steps in the validation process. As mentioned above, the earliest and simplest evaluation occurs during the proof-of-concept stage, where an early prototype of a technology designed for a medical environment is evaluated through a usability study, in both a laboratory setting and in a clinical environment where it is meant to be used. The next step, according to regulations in many countries regarding medical devices and technologies, is a clinical trial. Both of these are discussed in the sections below.

12.1   USABILITY TESTING

A usability evaluation involves presenting a prototype to users that simulate some of the aspects of system functionalities that will eventually make-up the final product. Through visual inspection of the interface, most users gain an impression of how they will react to the system when it has been fully implemented. Three formal usability study methods are briefly presented below: a heuristic evaluation, a cognitive walkthrough, and a videotaped evaluation.

12.1.1   HEURISTIC EVALUATION

A heuristic evaluation is an inspection method for computer software that helps to identify problems that may lie in the user-interface (UI) part of the design. Evaluators examine the interface and judge its compliance with recognized usability principles (the "heuristics"). Heuristic evaluations are one of the most informal methods of usability inspection in the field of human-computer interaction (HCI) and most useful in the early stages of development of a new software tool (Nielsen and Molich, 1990; Wikipedia_heuristics).

12.1.2 COGNITIVE WALKTHROUGH

A cognitive walkthrough begins with the analysis of a task, with a specific sequence of steps or actions required by a user to accomplish the task, and system responses to those actions. The designers and developers of the software then walk through the steps as a group, asking themselves a set of questions at each step. Data is gathered during the walkthrough and potential issues are recorded. Then the software is redesigned to address the issues which were identified. The effectiveness of methods such as cognitive walkthroughs is hard to measure in applied settings, as there is a limited opportunity for controlled experiments while developing the software (Wharton et al., 1994).

12.1.3 VIDEOTAPED EVALUATION

As the name implies, in a videotaped evaluation, the recording system replaces the observer and the users are recorded while they perform the requested tasks. Several methods can be used to perform this type of evaluation, including the use of questionnaires, interviews, focus groups, performance measurements, thinking aloud protocols, field testing, and user logs. In the case of the 'think-aloud' approach, users are asked to speak aloud about their thoughts as they do each of the tasks; these comments are noted by the observer. This is an approach which works well in a videotaped session (Rubin and Chisnell, 2008; Weyand, 2011).

Many aspects of usability can be studied by simply allowing the user to try the new tool or device and record their comments and feedback. Since CDSSs are highly dependent on user participation, the research team must exercise care in collecting and analyzing the users' feedback. A first step is to assess the need for the new tool proposed to potential users, and the goals of the usability study to evaluate the tool need to be clear. As the prototype goes through a number of updates, users should be asked to assess it at each substantial step to ensure the design is acceptable to them. Comments and feedback should be collected at each stage of the design and testing process.

Usability testing with participants who are most likely to be the future users is very important throughout the development of a computerized CDSS. Usability testing aims to assess the usefulness of a new device or software tool; its efficiency, how many steps and how much time is needed to complete all the tasks; accuracy, that is how many mistakes are made and whether they are recoverable or not; its effectiveness, that is, does it do what it is intended to do; and user satisfaction, that is whether users are confident or stressed while using the tool and would they consider using it in their workflow (Nielsen, 1994; Wikipedia_usability; Frize et al., 2005; Weyand, 2011). These factors are critical for a successful implementation.

The measurements will depend on what usability approach is chosen. One approach is the Goal-Question-Metric (GQM) method developed in the 1980s by Basili. This approach links the questions to be asked to the goals to be achieved with the testing. It is common to use a Likert scale for these studies, either with four or five choices; five choices would normally be: Strongly agree, agree, neutral, disagree, and strongly disagree. In the case of four choices, the neutral response is removed, and this may help to avoid what is called a central tendency, where persons choose the neutral response to avoid having to agree or to disagree. To obtain a quantitative result for the testing, a metric is assigned to each question; for example, a threshold can be set at 70%; if answers of 'agree' and 'strongly agree' are combined and make up 70% or more of the total number of responses, then the conclusion would be that the question met the criteria for success (Basili, 1984; Weyand, 2011). Measurements should include the time to complete each task, the number of errors and mis-clicks, and of course all comments made during the test.

Finally, the study should determine how many participants will be invited to evaluate the tool. The literature mentions using between 4 and 12 participants and that there is no added value to having more than 12 (Weyand, 2011; Lewis 2012).

12.2 CLINICAL STUDIES AND TRIALS

The U.S. National Institutes of Health (NIH) defines a clinical study as research that involves human volunteers (participants) with the intention to add to medical knowledge. There are two main types of clinical studies: observational studies and clinical trials.

In an **observational study**, investigators assess health outcomes in groups of participants according to a protocol or research plan. Participants may receive interventions, which can include medical products, such as drugs or devices, or procedures as part of their routine medical care, but participants are not assigned to specific interventions by the investigator (as in a clinical trial). For example, investigators may observe a group of older adults to learn more about the effects of different lifestyles on cardiac health (**NIH_Observational studies**).

In a **clinical trial**, participants receive specific interventions according a research protocol. These interventions may be medical products, such as drugs or devices; procedures; or changes to participants' behavior. Clinical trials may compare a new medical approach to a standard one that is already available or to a placebo that contains no active ingredients and no intervention. Some clinical trials compare interventions that are already available or to a placebo that contains no active ingredients or to no interaction (**NIH_Clinical trials**).

It is important to study any new medical intervention or a modification of a procedure to assess if it is safe and effective and whether it the same, better, or worse than other available approaches. An example would be for investigators to prescribe a new medication that is expected to lower blood pressure and test it against current medication used for that effect. To avoid any potential bias, the researchers can use either the single blind or the double blind approach. In both cases, the participants are divided into two groups: one group is to be administered the new intervention and the other group, being a control group, will either receive a placebo or the currently accepted intervention. For single blind studies, either the participants or the researcher do not know which participant is receiving the new intervention to be tested and which participant is in the control group. For double blind studies, neither the participant nor the researcher know how the participants are divided into the two groups.

Prior to doing any study with humans, the researchers need to apply to an Ethics Review Board at the institution where the research is being done to obtain a clearance or ethics certificate. Researchers also need to apply the four fundamental ethics principles to their study to ensure that they respect the participants' autonomy, that the experiments will be beneficial to humans, they will respect non maleficence (doing no harm), and the principle of justice, that is, to benefit as much of the population as they can (Frize, 2011).

The application process may vary from one institution to the other, but generally the following will be needed (see Frize, 2011) for more details:

- The submission of a protocol with goal, objectives, background, methodology, analysis of results, and a statement on the sample size needed as justified by a statistical analysis.

- Inclusion and exclusion criteria: what factors will exclude participants and which ones will be considered positive for enrolment in the study.

- The procedure for recruitment of participants; a direct relationship between participant and researcher must be avoided; for example a patient and the physician treating him or her. The advertisement poster and letter must be included in the request for clearance.

- The questionnaire, if there is one.

- A description of any device to be used.

- Any potential harm that can happen.

- A consent form that explains clearly in terms understood by someone with an eighth-grade education, the purpose, what will be done, how long the experiment is expected to last (minutes or hours, and number of times the participant is expected to be subjected to it).

- Potential benefits of the study, either for the participants or for people in general; usually there is no direct benefit for the participants but there can be benefits in the future for either a specific population or for the population at large.

In closing this chapter, it is important to note that using a placebo with the control group in a study may not be ethical if it means these participants are not treated appropriately for a medical condition that needs a treatment. What is more acceptable today would be to compare a new approach or medication with the best existing one at the time of the study. The results can then identify whether the new approach or medication works better, is equal to, or works less well than the currently recommended technique. Placebos can be used when no harm is done to participants by withholding a treatment.

REFERENCES

Basili, VR, Weiss, DM (1984). "A Methodology for Collecting Valid Software Engineering Data." *IEEE Trans. Software Eng.*, SE-10: 728-738. DOI: 10.1109/TSE.1984.5010301.

Faulkner, L (2003). "Beyond the Five-User Assumption: Benefits of Increased Sample Sizes in Usability Testing. Behavior Research Methods, Instruments." *Computers*, vol. 35: 379-383.

Frize, M, Yang, L, Walker, RC, O'Connor, AM (2005). "Conceptual Framework of Knowledge Management for Ethical Decision-Making Support in Neonatal Intensive Care," *IEEE Trans. Inf. Tech. Biomed.*, 9: 205-215. DOI: 10.1109/TITB.2005.847187.

Frize, M (2011). *Ethics for Bioengineers*. Morgan & Claypool. Synthesis Lectures on Biomedical Engineering. Series ed. J. Enderle. DOI: 10.2200/S00393ED1V01Y201111BME042.

Hussain, A, Ferneley, E (2008). "Usability Metric for Mobile Application: A Goal Question Metric (GQM) Approach." *Proc. Intern. Conf. on Information Integration and Web-based Applications and Services*: 567-570. DOI: 10.1145/1497308.1497412.

Lewis, JR (2012). "Sample Sizes for Usability Studies: Additional Considerations." *J. Bus. Tech. Commun.*, 26: 171-201.

Nielsen, J (1994). Usability Engineering. San Diego: Academic Press. pp. 115–148.

Nielsen, J, and Molich, R (1990). "Heuristic Evaluation of User Interfaces." *Proc. ACM CHI'90 Conf.* (Seattle, WA, 1–5 April), 249-256. DOI: 10.1145/97243.97281.

NIH_Observational studies (n.d.). Available at: http://clinicaltrials-lhc.nlm.nih.gov/ct2/about-studies/learn#ObservationalStudies; last accessed September 2013.

NIH_Clinical trials (n.d.). Available at: http://clinicaltrials-lhc.nlm.nih.gov/ct2/about-studies/learn#ClinicalTrials; last accessed September 2013.

Rubin, J, Chisnell, D (2008). *Handbook of Usability Testing: How to Plan, Design, and Conduct Effective Test.* Wiley Publishing Inc. Indianapolis, IN.

Solingen, RV, Berghout, E (1999). *Goal/Question/Metric Method: A Practical Guide for Quality Improvement of Software Development.* McGraw-Hill Education.

Weyand, SA (2011). "Development and Usability Testing of a Neonatal Intensive Care Unit Physician-Parent Decision-Support Tool." MASc Thesis in Biomedical Engineering. Carleton University, Ottawa, Canada.

Weyand, SA, Frize, M, Bariciak, E, Dunn, S (2011). "Development and Usability Testing of a Combined Physician-Parent Decision-Support Tool for the Neonatal Intensive Care Unit." *Proc. IEEE, EMBC,* Boston, Sept: 6430-6433.

Wikipedia_heuristics. Available at: http://en.wikipedia.org/wiki/Heuristic_evaluation; last accessed July 2013.

Wharton, C, Rieman, J, Lewis, C, Polson, P (1994). "The Cognitive Walkthrough Method: A Practitioner's Guide." In J. Nielsen & R. Mack. *Usability Inspection Methods.* John Wiley and Sons, NY: 105-140.

Wikipedia_usability. Available at: http://en.wikipedia.org/wiki/Usability; last accessed November 2013.

OTHER SUGGESTED READING

Friedman, LM, Furberg, CD, DeMets, DL (2010). *Fundamentals of Clinical Trials.* 4th Ed. Springer. DOI: 10.1007/978-1-4419-1586-3.

Frize, Monique (2011). *Ethics for Bioengineers.* Synthesis lectures on Biomedical Engineering, Morgan & Claypool Publishers. DOI: 10.2200/S00393ED1V01Y201111BME042.

Hornbaek, K (2006). "Current Practice in Measuring Usability: Challenges to Usability Studies and Research." *Intern. J. Comp. Studies,* 64(2): 79-102. DOI: 10.1016/j.ijhcs.2005.06.002.

Jacobsen, NE, Hertzum, M, John, BE (1998). "The Evaluator Effect in Usability Studies: Problem Detection and Severity Judgment." *Proc. Human Factor & Ergonomics Soc. Annual Meeting.* 42(19): 1336-1340. DOI: 10.1145/286498.286737.

Kunert, T (2009). *User-Centered Interaction Design Patterns for Interactive Digital Television Applications* (Human-Computer Interaction Series). Springer. DOI: 10.1007/978-1-84882-275-7.

Molich, R, Nielsen, J (1990). "Improving a Human-Computer Dialogue." *Commun. ACM,* 33, 3 (March), 338-348. DOI: 10.1145/77481.77486.

Nesbitt, LA (2004). *Clinical Research: What it is and How it Works.* Jones and Bartlett Publishers.

Nielsen, J (1994). "Heuristic evaluation." In Nielsen, J., and Mack, R.L. (Eds.), *Usability Inspection Methods.* John Wiley & Sons, New York, NY

Nielsen, J (1994). *Usability Engineering.* San Diego: Academic Press. pp. 115–148.

Pocock, SJ (1983). *Clinical Trials: A Practical Approach.* John Wiley

Author Biography

Monique Frize is a Distinguished Professor at Carleton and Professor Emerita at University of Ottawa. For 18 years (1971-1989) she was a hospital biomedical engineer and has been a professor in electrical and biomedical engineering since 1989. Monique Frize has published over 200 journal and conference proceedings papers on artificial intelligence in medicine, infrared imaging, ethics, and women in engineering and science. She is a Fellow of IEEE (2012), the Canadian Academy of Engineering (1992), Engineers Canada (2010), Officer of the Order of Canada (1993), and recipient of the 2010 Gold Medal from Professional Engineers Ontario and the Ontario Society of Professional Engineers. She has received five honorary doctorates in Canadian universities since 1992. Her book, *The Bold and the Brave: A History of Women in Science and Engineering* was released by the University of Ottawa Press in November 2009; *Ethics for Bioengineers* was published by Morgan & Claypool (2011); and her new book, *Laura Bassi and Science in 18th Century Europe: The Extraordinary Life and Role of Italy's Pioneering Female Professor*, was released by Springer in July 2013.

CPSIA information can be obtained at www.ICGtesting.com
Printed in the USA
BVOW06s1916090114

341215BV00002B/2/P